Engineering Graphic Modelling

Engineering Graphic Modelling

A Workbook for Design Engineers

E. Tjalve
M. M. Andreasen
F. Frackmann Schmidt
*(Laboratory of Engineering Design,
The Technical University of Denmark)*

English translation edited by Dr. Geoff Pitts,
*(Department of Mechanical Engineering,
Southampton University)*

NEWNES–BUTTERWORTHS
LONDON · BOSTON
Sydney · Wellington · Durban · Toronto

The Butterworth Group

United Kingdom **Butterworth & Co (Publishers) Ltd**
London: 88 Kingsway, WC2B 6AB

Australia **Butterworths Pty Ltd**
Sydney: 586 Pacific Highway, Chatswood, NSW 2067
Also at Melbourne, Brisbane, Adelaide and Perth

Canada **Butterworth & Co (Canada) Ltd**
Toronto: 2265 Midland Avenue, Scarborough, Ontario, M1P 4S1

New Zealand **Butterworths of New Zealand Ltd**
Wellington: T & W Young Building, 77–85 Customhouse Quay, 1, CPO Box 472

South Africa **Butterworth & Co (South Africa) (Pty) Ltd**
Durban: 152–154 Gale Street

USA **Butterworth (Publishers) Inc**
Boston: 10 Tower Office Park, Woburn, Mass. 01801

First published in Danish as 'Grafiske Modeller' 1976
First published in English 1979

British Library Cataloguing in Publication Data

Tjalve, E
Engineering graphic modelling
1. Engineering graphics 2. Engineering design
I. Andreasen, M Myrup II. Schmidt F Frackman
604'.2'4 T353 78-40582

ISBN 0-408-00305-7

Typeset by Butterworths Litho Preparation Department
Printed in England by Camelot Press Ltd, Southampton

Preface

Our aim in writing this book is to bring together the basic material which we have evolved during the past years for the teaching of mechanical engineering design draughtsmanship. We have adopted what is, in many ways, a novel approach, which aims to develop the student's grasp of three-dimensional perception with the creative use of freehand sketching, which is the essential mode of expression in conceptual design.

The book is made up of individual work sheets arranged in a logical order. The method of using the book is fully explained on page 1, and the student or draughtsman is urged to read this carefully before delving into the book. Once understood in their inter-relationship, these work sheets will be of continuing value for reference purposes. We must emphasise that the book is essentially an introduction to draughtsmanship in a similar way to which a book of exercises and scales is to the young musician.

The book aims to develop a fluency of thought and expression, through a mastery of the manual skills which play an all-important role in design draughtsmanship. For a grounding in formal drawing office practice, the student should refer to one of the many texts which are available.

When an established specialist subject is treated from a basically novel point of view, problems of terminology will arise. These problems are magnified in translation into another language. We are very grateful to Dr. Geoff Pitts for his work in bringing the book into accordance with English drawing practice and terminology.

E. Tjalve
M. M. Andreasen
F. Frackmann Schmidt

Editor's Foreword

It is a pleasure to have been invited to edit the English translation of this work. The text provides an unusual approach to the topic of engineering drawing. Often the authors of such books become preoccupied with setting out the rules at the expense of relating the subject to the broader engineering function.

The authors of this text have set out to show how engineering drawing relates to the design activity, and by so doing should stimulate the interest of the student in what can become a very stereotyped subject. The book should also cause the designer and draughtsman to think more carefully about the role of drawing in engineering design communication.

A number of changes have been made to the original version to relate it to the English speaking reader. Conventions in drawing have evolved on the basis of what has been satisfactory in use. Some conventions are obvious because they are a pictorial representation resembling the actual situation, or because they are sufficiently close to a standard convention as to be interpreted accurately, the latter situation often occurs where a company uses its own version of a national standard. For these reasons it will be found that not all the examples rigorously follow the national or international drawing conventions, to do so would limit the scope of the book in its attempt to project drawing as a communication medium in which the creator and the receiver have knowledge of the drawing code.

Finally, it is hoped that the changes which have been made will not detract from the excellence of the original volume.

Geoffrey Pitts

Contents

How to Use this Book

Every stage of engineering design requires some form of drawing to support it. The type of drawing depends on the purpose of the drawing and the information it contains. This book sets out the different drawing types and relates them to the situations in which they are used.

It is important that the following introduction on pages 2–12 is studied first, because it provides the thread linking the various sections of the text. The drawing key on page 13 can then be used to identify the work sheet of interest.

TERMINOLOGY
In order to communicate basic concepts the authors have used some words in their 'purist' form, which is not necessarily the meaning normally applied to them by engineers. The most obvious examples are:

Structure being the manner in which ideas *OR* concepts are brought together *OR* the way in which system elements are arranged.

Surface being the general concept of expressing the properties of the surface and more specifically surface finish.

DRAWING AND DESIGN

DRAWING - FROM AN ENGINEER'S VIEWPOINT

Designing may be understood as devising and making machines and products to satisfy human needs. The process, leading from the *need* to the finalised and approved drawings and specification is called the design process. In this design process the designer uses a number of 'aids'. Their type depends on the product in question and the stage in the design process which has been reached. *Drawing* is the most important aid to the design engineer.

A drawing is a universal means of modelling an idea or a proposed solution, both in the abstract and at detail design level. It enables the designer (and others) to examine the consequences of his ideas and to evaluate the properties of possible solutions. The completed product is influenced to a high degree by a designer's ability to draw.

Traditionally, a mechanical engineer's drawing activities and sometimes the whole activity of design have been limited to the preparation of working drawings e.g., layouts, detail and assembly drawings. But why produce the final drawings of a product, if the designer has been unable to formulate and modify his ideas during earlier creative stages? Here, rapid and imaginative drawing and sketching techniques are fundamental to the final result. It can thus be concluded that each phase in the design process requires different drawing skills from the engineer.

WHAT IS A DRAWING?

Two different aspects illustrate the real purposes of drawing.

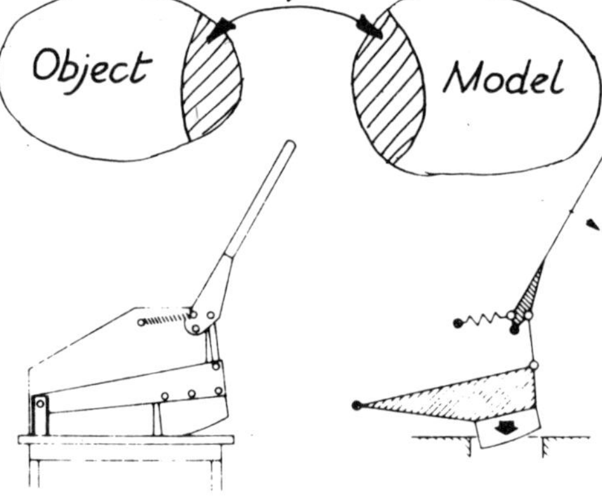

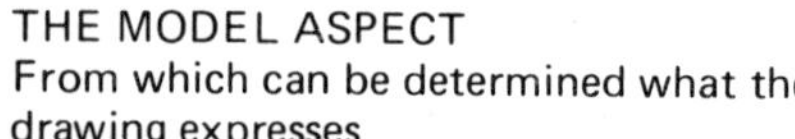

THE MODEL ASPECT
From which can be determined what the drawing expresses.

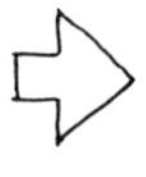

THE COMMUNICATION ASPECT
From which can be obtained a general idea of the drawing's use.

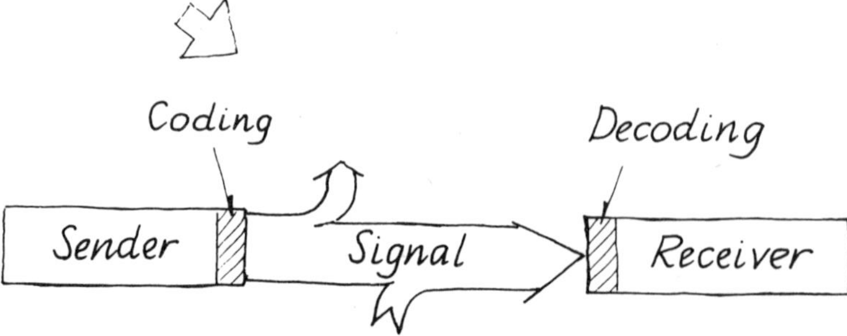

A drawing always models something (i.e. an item *OR* a system). This can be called the OBJECT. The model has some characteristics in common with the object. A model may be described by the following features:

TYPE OF MODEL
e.g. mathematical model, model of form, detail drawing

MODELLED PROPERTIES
e.g. function, structure, form

USE
e.g. simulation, verification, investigation.

A model may represent only some of the properties of the object and differs in this respect from the object, e.g. a functional model has the property function in common with the object whilst, for example, the form is different. Similarly, the model may contain properties not found in the object. The form of a functioning model may be different from that of the object: materials, friction and sensitivity to temperature may differ.

The advantage of the model is in being able to limit the model to those properties of interest, whilst omitting the remainder. In this way, specific properties can be examined quickly and cheaply. Its weakness is in it having properties not existing in the object, which may have an unintentional effect.

When modelling, or drawing, it is important to be conscious of differences, so that the most appropriate model for the situation is chosen.

Communication by drawing may either be with others or be a means for the designer to reflect his ideas. The drawing communicates information about some of the objects' properties.

Information passes from sender to receiver. The information is coded into a signal: the signal then becomes a model of the information. This is decoded by the receiver, and only then has the receiver obtained the information. The code e.g. words, electrical impulses or symbols, must be known to the sender and receiver.

The information transmitted may be added to or lost if the receiver does not perceive the whole signal or if there is an input from another source.

A drawing used for communication is a signal (coded information), and therefore three requirements must be met for the information transfer to be successful.

The RECEIVER must be known (USE).
The CODE must be appropriate to the receiver.
The required quality of coding must be known, i.e. which drawing technique should be used.

In addition the following must be determined:

DRAWING TOOLS (pencil, ball point, drawing pen).
DRAWING MEDIUM (tracing paper, sketching paper, projection transparency).

ANY DRAWING HAS 6 FEATURES

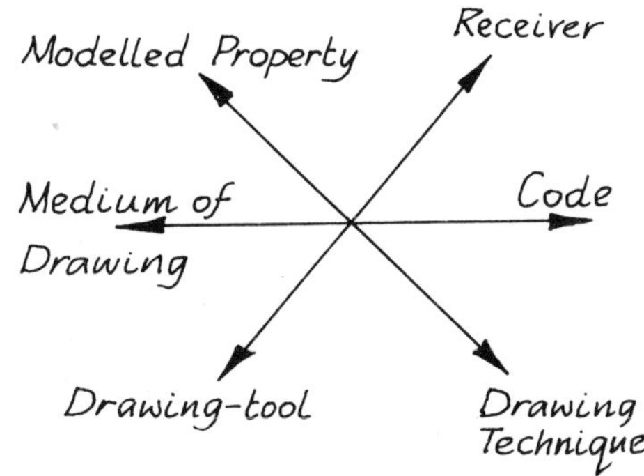

THE MORPHOLOGY OF DRAWING

Morphology: *The splitting up of something complex into characteristic elements*

A closer analysis of the listed features can be made by considering a particular activity. In the case of drawing for engineering design, a MORPHOLOGY OF DRAWING can be tabulated as shown on the next page. This shows the four important characteristics: modelled properties, receiver, code and the drawing technique, with a list of possibilities contained in each. On the extreme right of the page is a list of common drawing types, resulting from given combinations of the four characteristics.

Modelled properties

Any object (product, machine, system) possesses characteristic properties. Some of these properties are desired, but others may be more or less unwanted. Properties are manipulated throughout the design process. The first stage of the process is specifying the properties required of the finished product. Five properties dominate and together specify a product completely. These basic properties are: structure, form, material, dimension and surface. All other properties are dependent on these. A drawing can represent these properties directly, whilst other properties, of which function is the most important to an engineer, may be shown indirectly through the basic properties, or by giving the parametric connections. The way the various properties can be represented is given in the first group of work sheets.

Receiver

The receiver is the person or persons to whom the drawing communicates information. There are basically two different receivers: the designer himself (self communication) and others. The latter may be: other designers, technical draughtsmen, workshop staff, production planners, clients, managers and the public.

Code

Information transferred to a receiver through a drawing must be coded, i.e. translated into graphical symbols, which are known to the receiver. The codes may be split into three main categories: coordinates, symbols (machine parts, hydraulic components, electrical components, etc and those contained in standards for mechanical drawing), and types of projection.

Drawing technique

Technique is the method used to create the drawing, e.g. freehand sketch, use of a straight edge or draughting machine, use of templates. Generally, drawings can be produced by any of these techniques, but often accepted practice dictates a particular approach, e.g. detail drawings and patent drawings.

Drawing Morphology

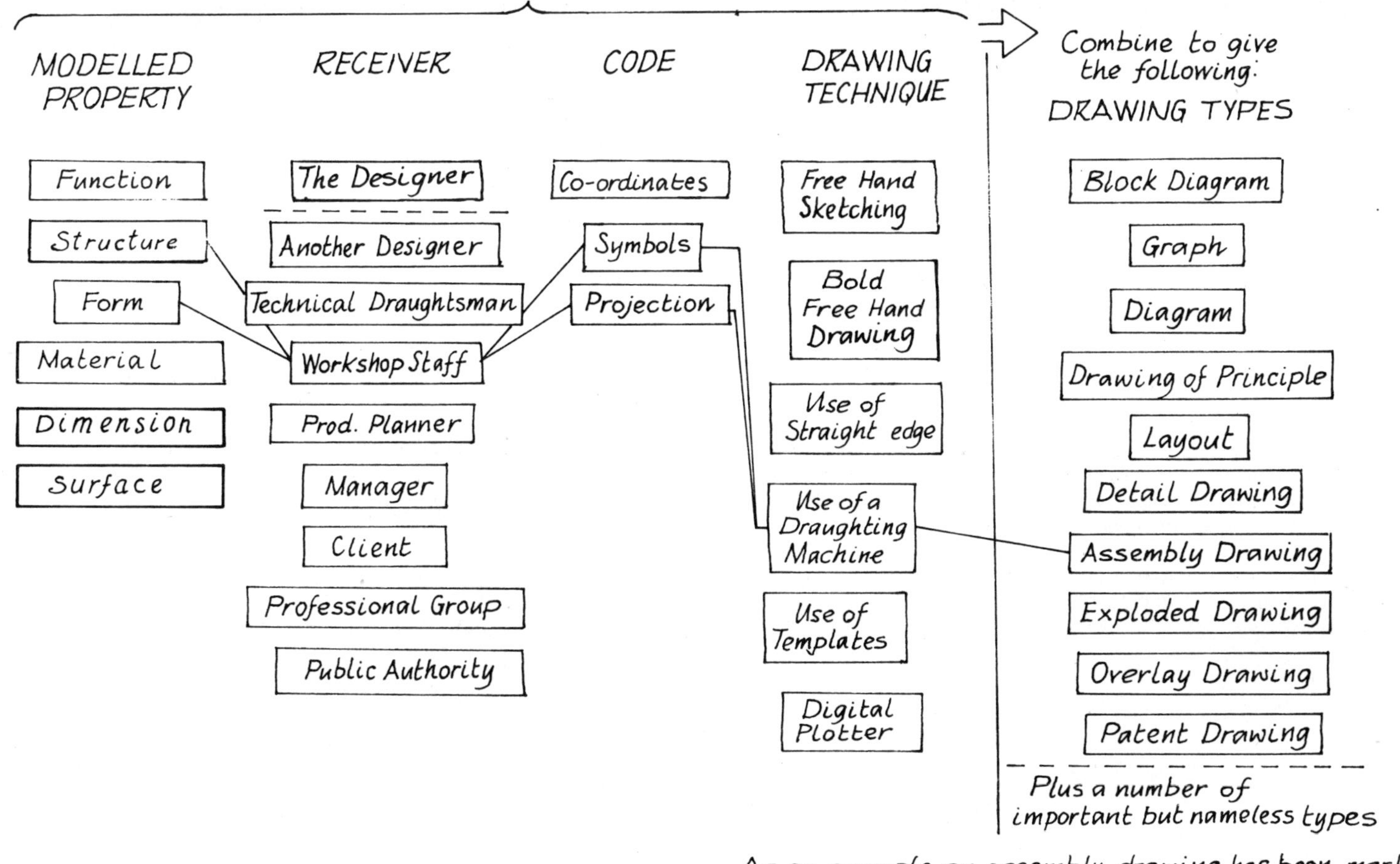

DESIGN

The purpose of design is to create products to satisfy human needs, the same process being followed regardless of the end product. Design within a specialised area demands specialist knowledge. An understanding of the interdependence of drawing and design results from a knowledge of the design process.

Problem solving may, in general, be represented by the following sequence of activities:

GENERAL PROBLEM SOLVING

Formulating the problem must result from a need, which is the sole purpose of designing. The designer's task is to decide the best route to satisfy the need, starting by analysing the problem, i.e. he must formulate the problem to which a design solution is sought.

Setting up the criteria involves defining the properties required of the solution and their relative importance. The criteria are the basis for selecting the most appropriate solution to the problem.

Seeking possible solutions is central to the activity. The aim is to identify a range of solutions using systematic and creative methods.

Investigating possible solutions may require drawings to examine the feasibility of the alternatives. These can become the embryo of the drawings for the selected solution. Calculations and experiments may also be ingredients at this stage.

Evaluation and selection results from a comparison between the properties required of the product and those offered by the solutions. The most appropriate solution is then developed further.

The design process is represented by this chain of activities followed by the designer to transform the acknowledged need into a mechanical system.

The product first emerges at the phase 'seeking solutions'. The starting point is satisfying the criteria specified in the analysis of the problem, which embrace the required function and properties.

TO DESIGN IS TO DEFINE THE BASIC PROPERTIES

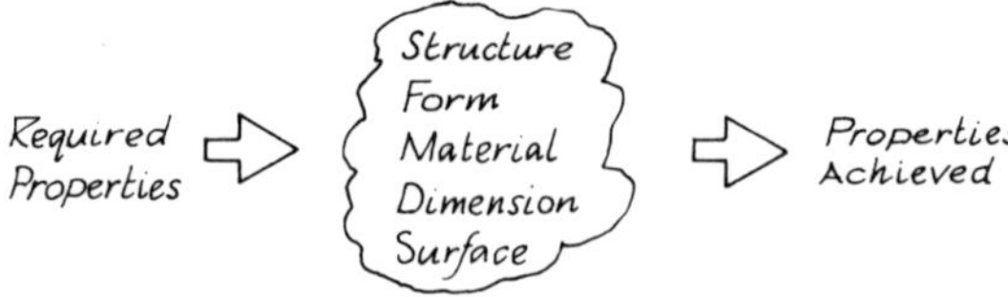

Required properties could be: ease of operation, pleasing appearance, reliability, good life expectancy, adjustment etc. These properties result indirectly from the basic properties: structure, form, material, dimensions and surface. It is the task of the designer to define the basic properties in such a way that the product acquires the required properties, see diagram.

The design process is better understood by seeing how the basic properties are established. Since the basic properties are interdependent, the sequence must be expressed as follows: first the structure is established; then, to a certain extent, the form of the elements; and finally the materials, dimensions and surface qualities together with the detailed form.

Product synthesis

is the procedure which, when followed, results in the properties required of the product.

Problem analysis
Main functions
Part functions and means
Basic structure
Quantified structure
Total Form
Form of elements Material Dimension surface

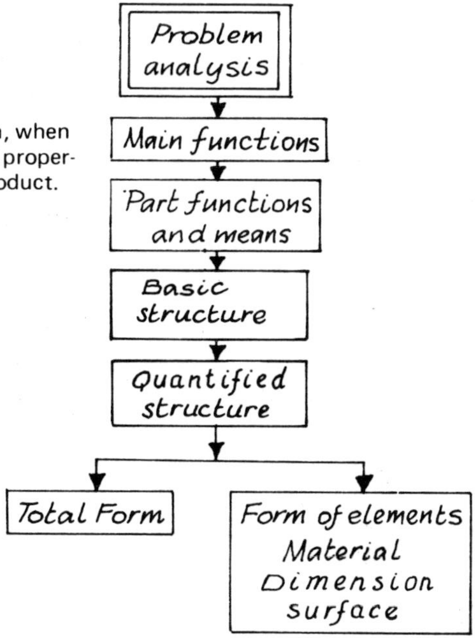

Design synthesis

is a step-by-step problem solving approach enabling the designer to organise his ideas. It's aim is to establish the basic properties, starting with the problem analysis to formulate the main function or functions (there may be several in parallel), required of the product.

The second stage is a division into sub-functions at various levels and the possible means of meeting them. After the possible means have been clarified, consideration of the structure can start. Two stages of structure follow, these occur at two levels. The form design follows the establishment of the structure and comprises two parallel activities: an establishment of the total form of the system, and establishing the initial form of the elements at a concept level. Finally, the detailed form of the elements is decided involving the choice of materials, dimensions and surface qualities.

EXAMPLE: DESIGN OF A WALK-WAY FOR MEDICAL ASSESSMENT

On the following three pages, examples are shown of the types of drawings and sketches used in a design project, from the initial investigation to the finished apparatus. The apparatus measures the walking characteristics of a person fitted with an artificial limb. From the many possibilities, the solution chosen was the measurement of contact forces between the feet and the floor whilst walking. The examples illustrate the need for the different levels of drawing each time an alternative solution is examined.

DRAWINGS USED IN THE DESIGN OF A:-
Walk-Way

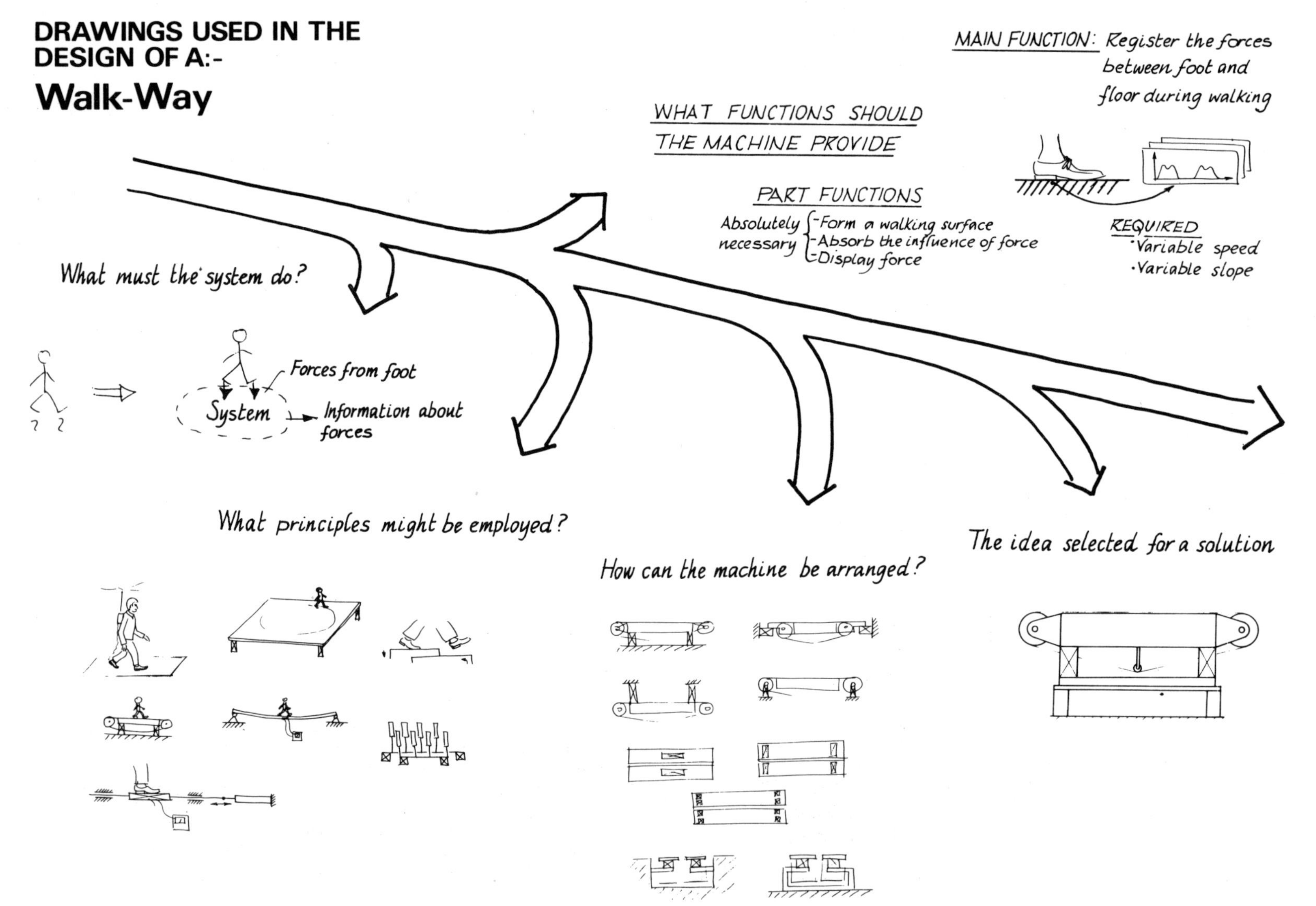

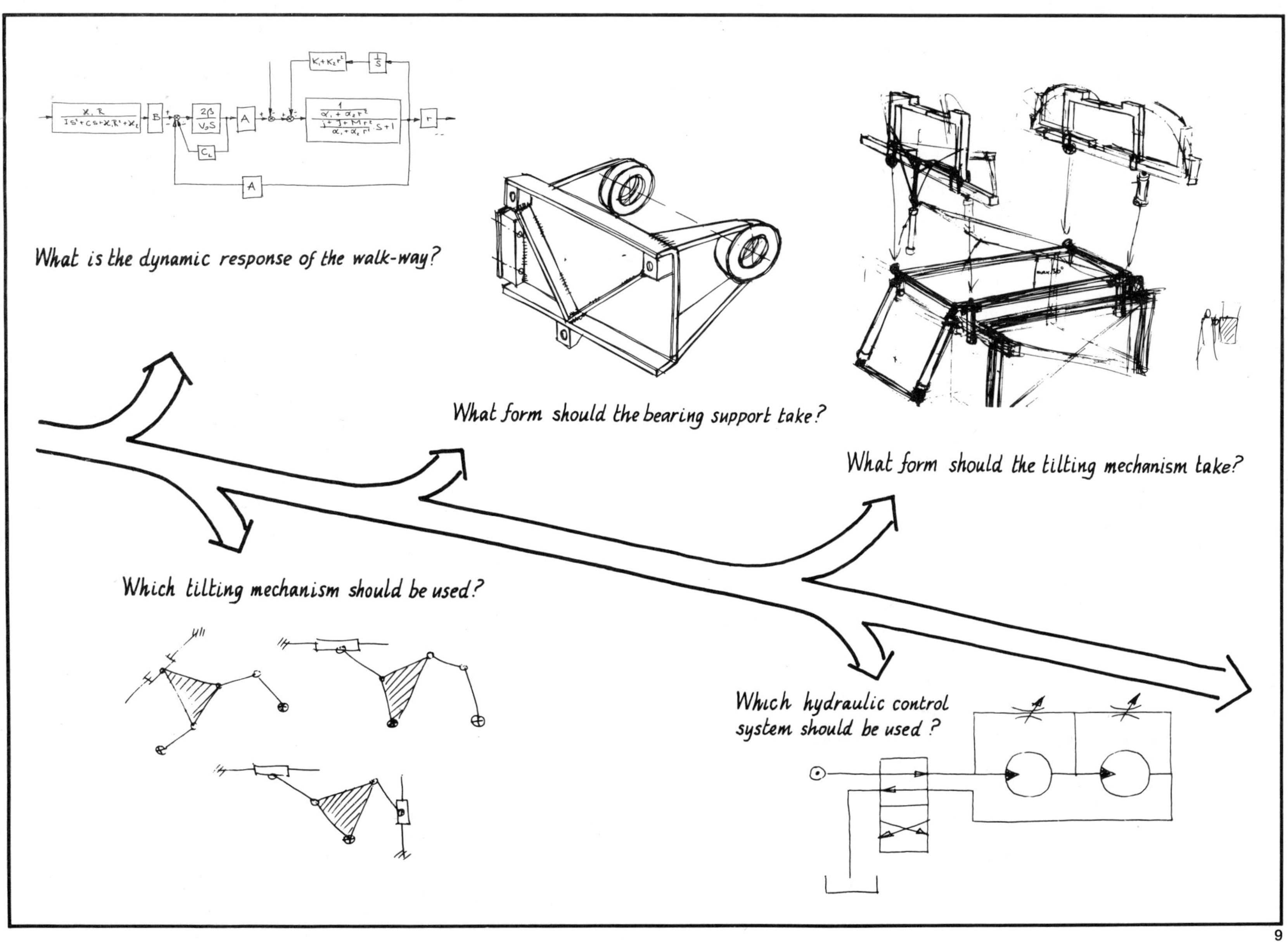
What is the dynamic response of the walk-way?
What form should the bearing support take?
What form should the tilting mechanism take?
Which tilting mechanism should be used?
Which hydraulic control system should be used?

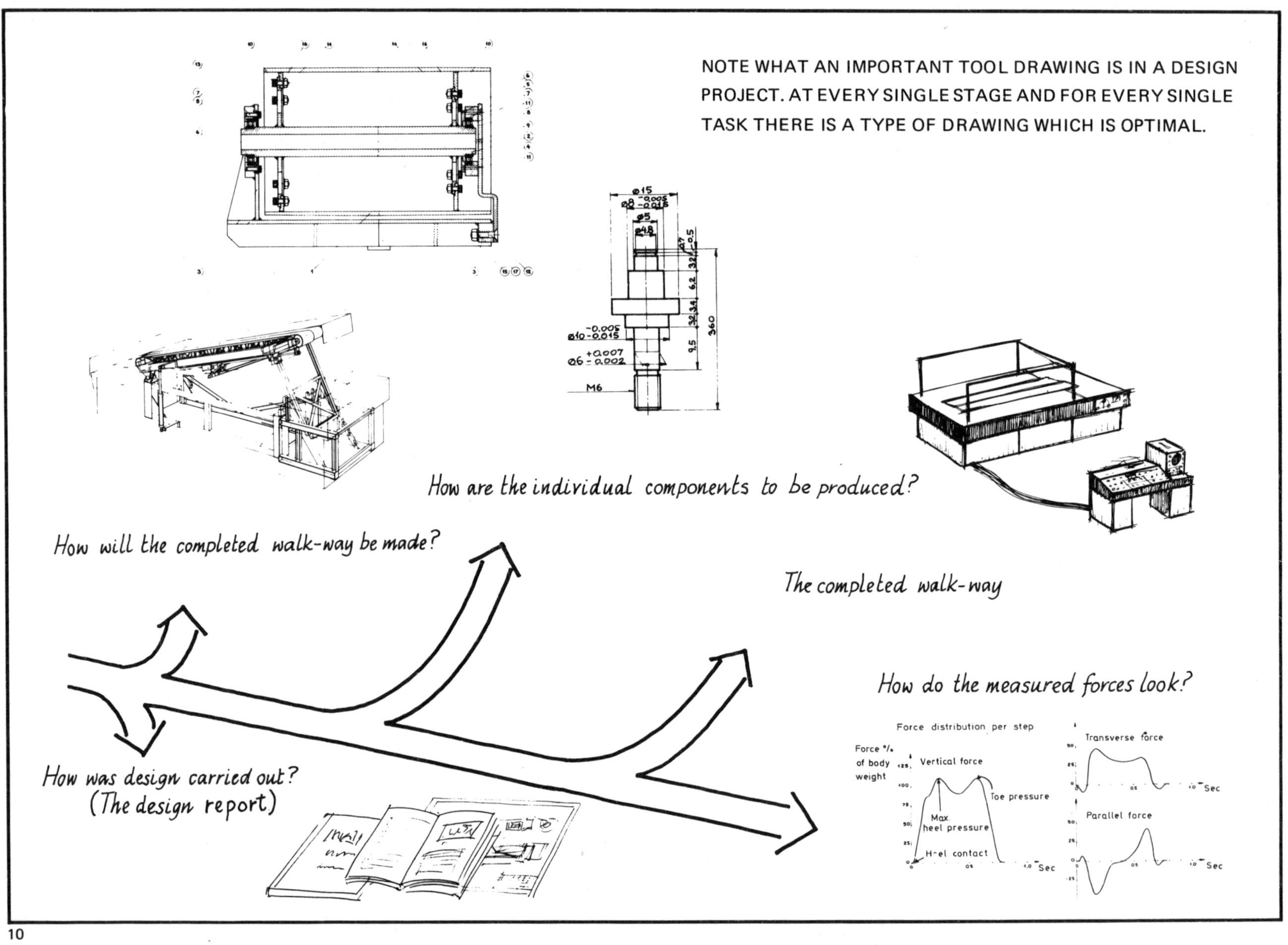
NOTE WHAT AN IMPORTANT TOOL DRAWING IS IN A DESIGN PROJECT. AT EVERY SINGLE STAGE AND FOR EVERY SINGLE TASK THERE IS A TYPE OF DRAWING WHICH IS OPTIMAL.
How are the individual components to be produced?
How will the completed walk-way be made?
The completed walk-way
How do the measured forces look?
How was design carried out?
(The design report)
Force distribution per step
Force % of body weight
Vertical force
Toe pressure
Max. heel pressure
Heel contact
Sec
Transverse force
Parallel force
M6
360

CHOICE OF DRAWING TYPE IN DESIGN

The morphology of drawing gives a general view of the drawing spectrum and consequently the characteristics of the invidual drawing types can be attributed, e.g. block diagram, detail drawing, and assembly drawing. It is evident that some combinations of the four characteristics are without name, such as: form – the designer – projection – freehand sketching, corresponding to an orthographic freehand sketch to aid the designer's own thinking. The list of drawing options is not exhaustive.

The starting point for a drawing is the designer's need to express specific properties in order to communicate them to himself or to the others; the code is decided from these two parameters i.e. the property and the receiver. The technique of drawing, the tools and the medium can then be chosen, giving the following progression:

MODELLED PROPERTY, RECEIVER } + CODE + DRAWING TECHNIQUE, DRAWING TOOL, DRAWING MEDIUM } ⇨ DRAWING

The selection of drawing type occurs in every design situation. In theory a choice is made on each occasion, in practice however, stereotypes emerge from characteristic situations. These traditions are expressed in substantially standardised drawing types e.g. detail drawings, assembly drawings, and therefore the process often starts with the drawing type as follows:

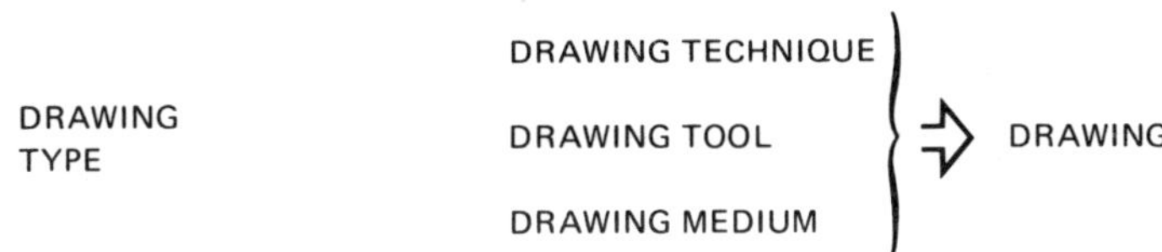

The two starting points represent the two extremes which in practice become merged. This usually results from the drawing types being poorly defined. It is important to realise that: modelled properties + receiver is the correct starting point to optimal drawing.

In general, only layouts, detail drawings, assembly drawings and patent drawings have well defined receivers and accepted drawing techniques, i.e. these types are characterised by definite situations of use. Creative work depends on self communication, and many of the drawings described, are of great importance in this context. The designer should therefore try to be objective in selecting the drawing types which support creative thinking at each phase of design.

Drawings should not be confined to stereotypes but, in each case, the property to be modelled, the recipient and the level of detail should be considered. By using the morphology of drawing the most appropriate type of drawing can be selected, which in some cases can yield unconventional results. The precise definition of the receiver, who will use the drawing, and the development of symbols will be considered here, resulting in purpose orientated drawings.

The following page sets out the morphology of drawing as an introductory key to the work sheets. Two starting points are shown providing two routes into the work sheets.

STARTING POINT 1

MODELLED PROPERTY + RECEIVER ⟶

Use sheets 1.1–1.6 to see how this property can be expressed.

Use sheet 2.0 to define the receiver's demands.

CHOOSE:

CODE + DRAWING TECHNIQUE ⇒ DRAWING

Use sheets 3.1–3.3 to study the applicable codes in the given situation. New codes can be defined in special situations.

Use sheets 4.1–4.6 to find a suitable drawing technique.

STARTING POINT 2

KNOWN:

DRAWING TYPE ⟶

Use sheets 5.1–5.17 to decide whether the chosen type of drawing is suitable for the given task.

CHOOSE:

DRAWING TECHNIQUE ⇒ DRAWING

Use sheets 4.1–4.6 to find a suitable drawing technique.

Arrangement of Work Sheets

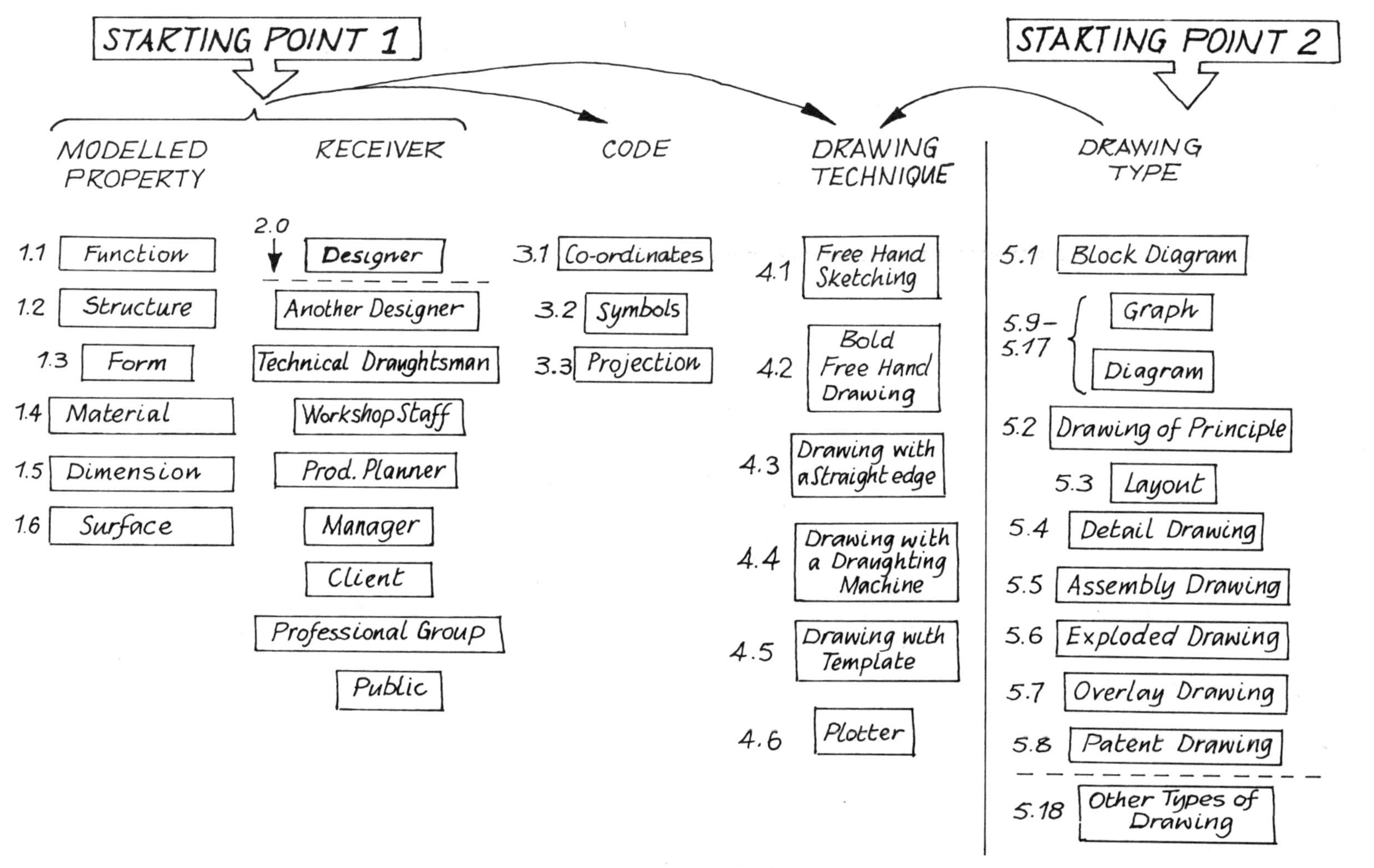

1. MODELLED PROPERTIES

1.0 SUMMARY

1.1 FUNCTION

1.2 STRUCTURE

1 3 FORM

1.4 MATERIALS

1.5 DIMENSION

1.6 SURFACE

1.0 MODELLED PROPERTIES

SUMMARY

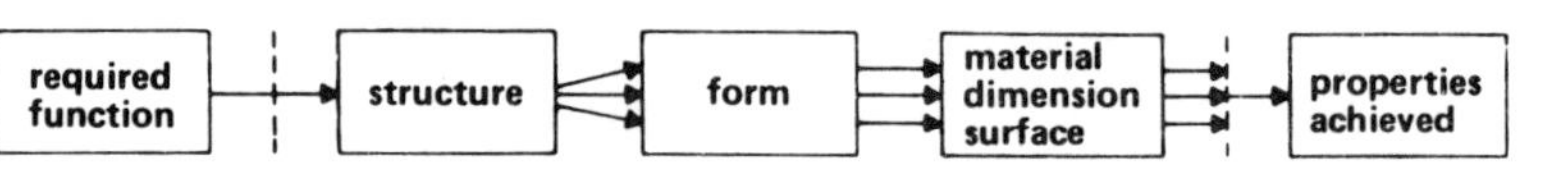

A property may be defined as a characteristic of an object, e.g. a machine, but this definition does not clarify precisely what the property is.

A machine, e.g. a screw-jack possesses a series of properties – shape, function, lifting capacity, use, reliability, resistance to corrosion, size, colour, etc. Here function is special because it characterises the machine, and justifies its existence i.e. the screw-jack can lift a weight.

Five other properties: structure, form, material, dimension and surface are special and are BASIC PROPERTIES. These properties are established as the design proceeds i.e. they are the variables which can be manipulated, and which decide all other properties.

It is important to distinguish between the two situations: (i) the design activity, where the properties are being established and (ii) the situation, where the properties are achieved or present in the product.

ESTABLISHING THE PROPERTIES

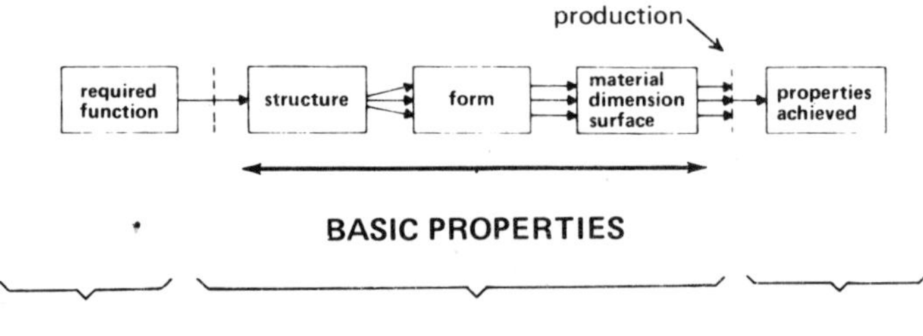

Modelling the basis of the design

Modelling as an aid to establishing the basic properties and also specifying for the production process

Modelling as a means of exploring the properties of the system achieved

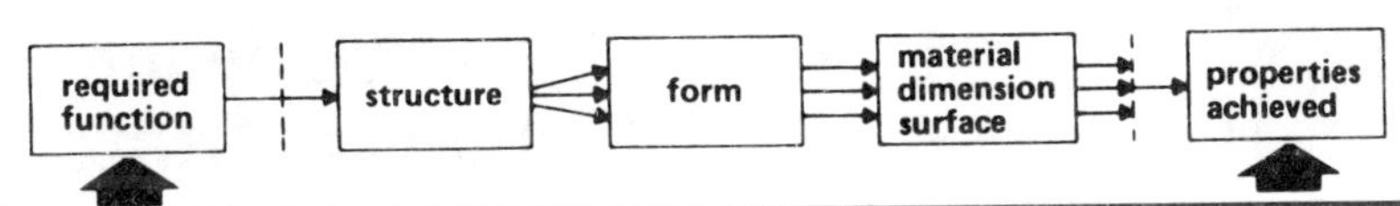

Definition

The function of a mechanical system is the transforming of the input to the system into the output from the system.

Use

The purpose of modelling FUNCTION is to examine how a given principle might be exploited as a possible solution to the problem. The function model can become a starting point for syntheses, resulting in a structure satisfying the desired function. The function should be described without using structural elements, as these may unnecessarily limit the choice in the search for a design solution. Descriptions of function are of interest at other stages, e.g. to tell the customer about the characteristics of the machine.

MODELLING THE DESIRED FUNCTION

1 MODELLING THE MECHANICAL STATES – representing the output states for a series of given input conditions.

2 MODELLING THE COURSE OF A PROCESS – a description of how the machine performs its function including changes and sequences in the machine.

3 MODELLING THE MAN/MACHINE-FUNCTION – an examination of the interrelationship between a machine and its operator.

4 MODELLING BY MEANS OF A STRUCTURE – an interpretation of the function from a study of the structure of physical elements.

1

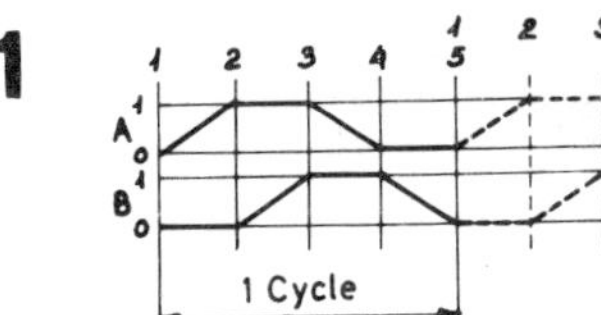

Modelling the mechanical states: this example shows the output from a system having two pneumatic cylinders.

2

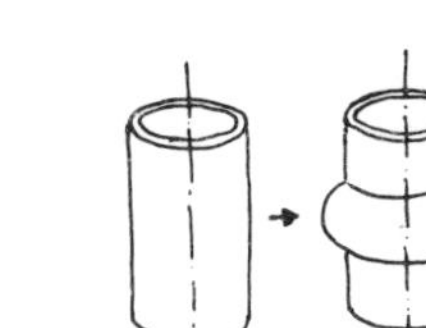

Modelling the course of a process: showing how the object is altered by the process.

3

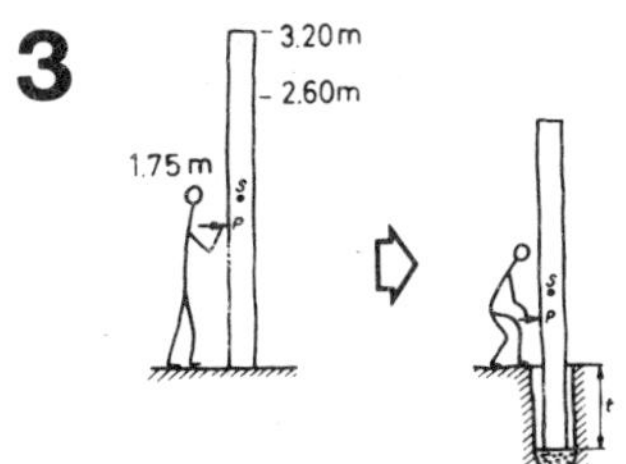

Modelling the MAN/MACHINE function: here the situation represents undefined equipment for putting a pole in the ground.

4

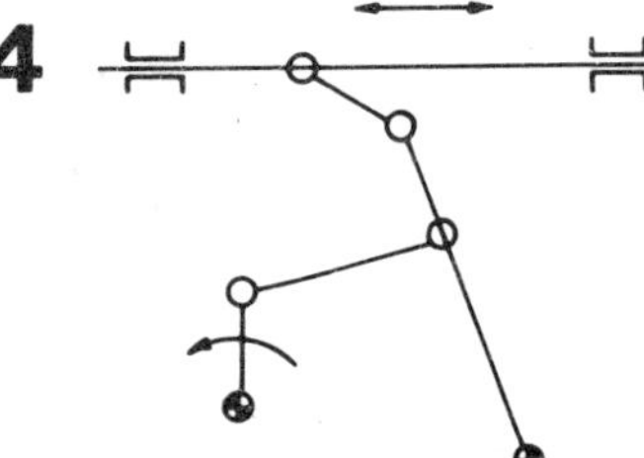

Modelling with the aid of a structure: by interpretation, the function resulting from the structure can be deduced.

FUNCTION 1.1

1	Modelling the mechanical states	2	Modelling the course of a process	3	Modelling of the man/machine function	4	Modelling by means of a structure

Modelling the Mechanical State

The condition of the machine can be modelled directly from its input/output. For a complex system, the function may be subdivided into blocks containing further statements of the function.

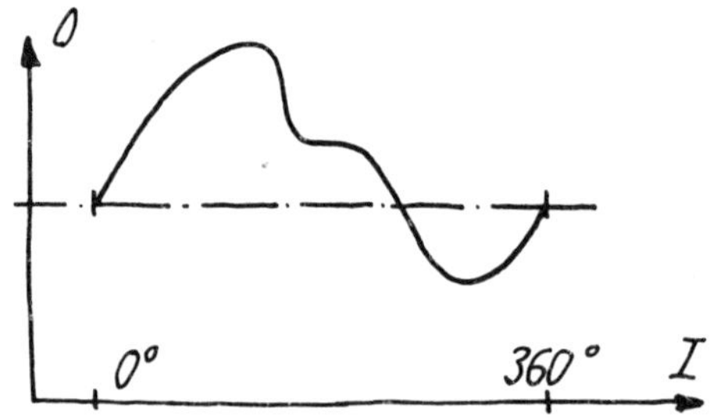

Curve showing the relationship between the rotating input to a mechanism and the required output.

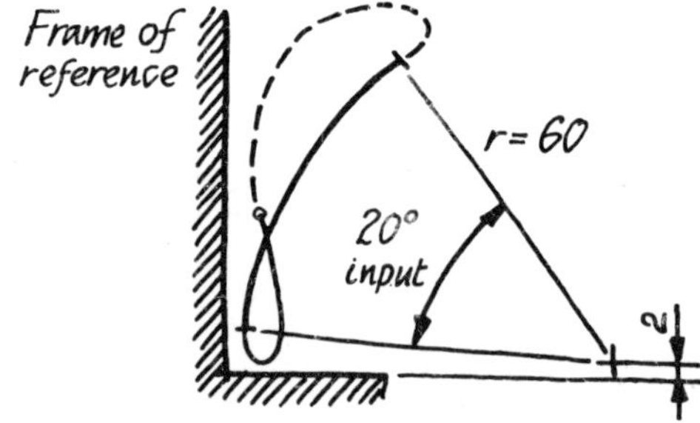

A sketched curve showing the output from an as yet unknown mechanism.

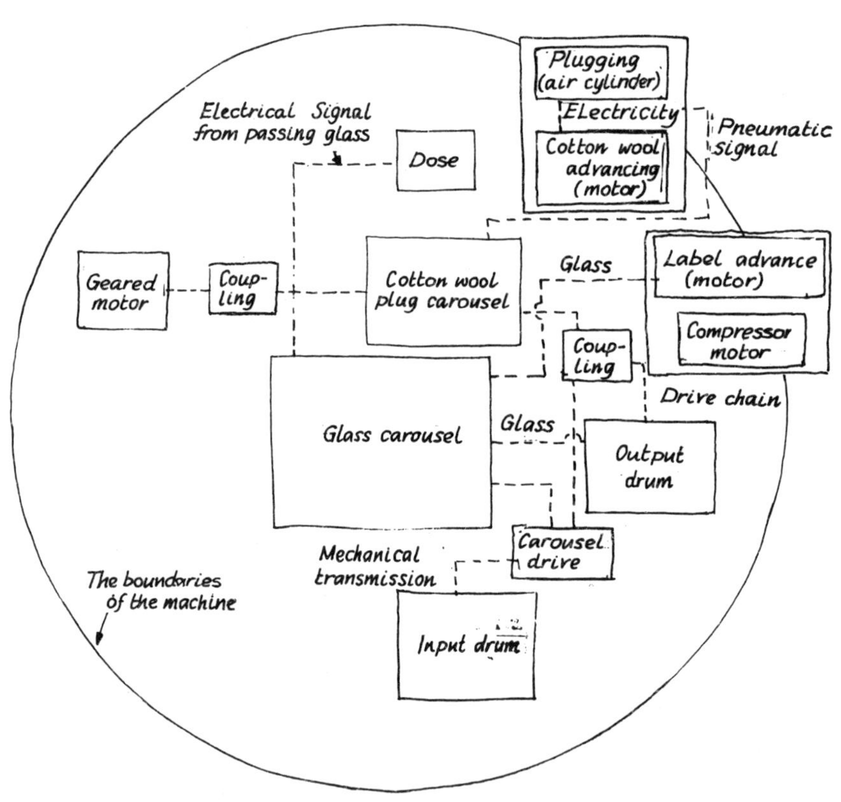

Block diagram showing the relationships between the subfunctions in an automatic test tube filling machine. The blocks have been arranged so that the total function is apparent.

Modelling the Course of a Process

The benefit from a mechanical system results from it performing a process. The process sets the objectives for the desired functions of the machine. Conversely a description of the process in turn describes the function.

This can occur in two ways:

A process is a progression in time, operating on an object, e.g. material, information, energy. A change of properties occurs, resulting from the mechanical system producing the desired changes.

An ELEMENT OF A PROCESS is a bounded change of properties simple or complicated, which can occur in a bounded PART of a machine SYSTEM.

MODELLING BY DESCRIBING THE ELEMENTS OF THE PROCESS

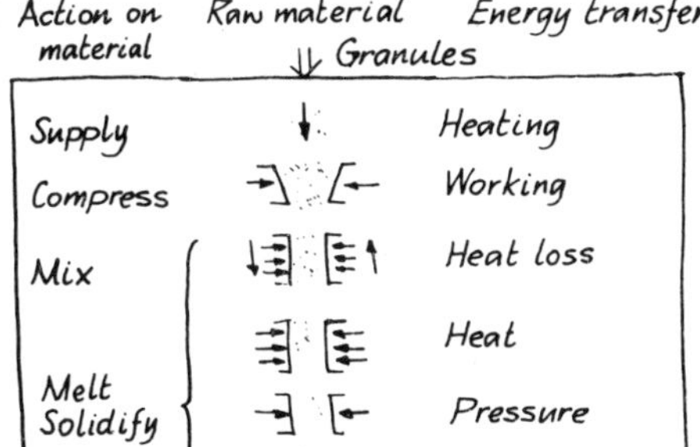

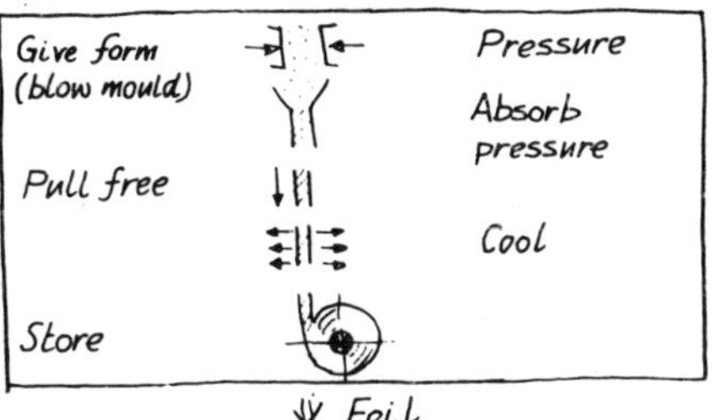

MODELLING BY DESCRIBING THE SUB-SYSTEMS

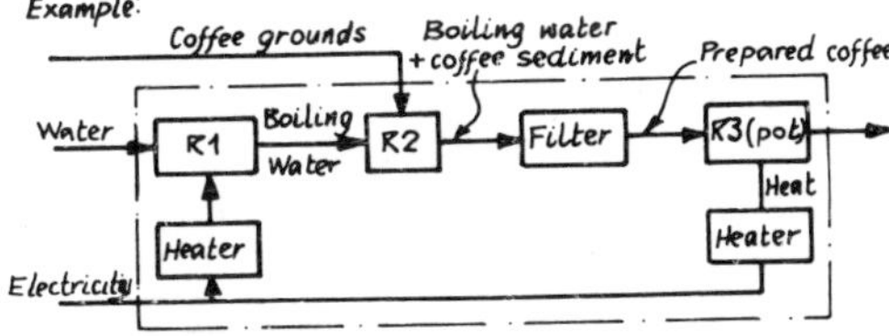

A block diagram describing a system. The arrows indicate the course of the processed object, and also the energy flow between the individual sub-system.

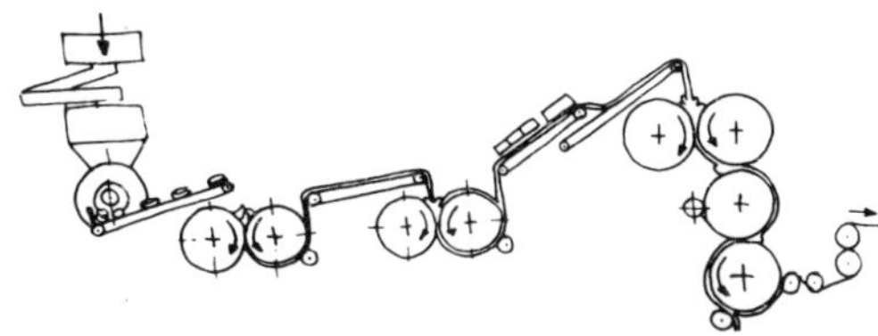

The sequence of a process implied by drawing the object of the process, and the arrangements of parts which comprise the process; in this case plastic sheet.

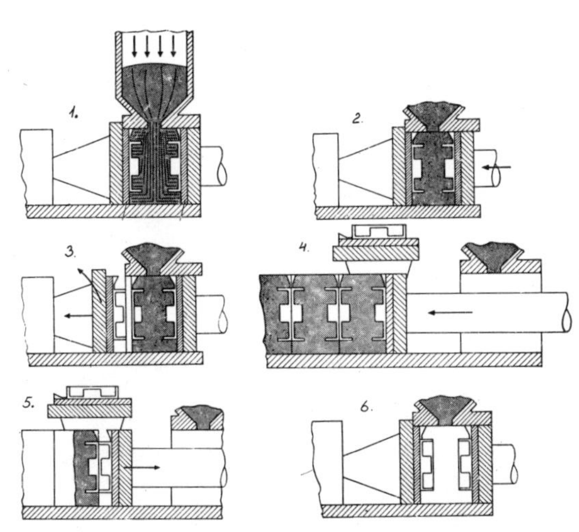

A diagrammatic way of describing the progress of sand through an automatic mould making machine; the sand being the object of the process. The function is shown in the series of pictures.

Modelling the Man/Machine Function

An operator's relationship to a task is analogous to the relationship between a mechanical system and a process. Both the operator and the mechanical system influence the object of the process, to produce the desired result. The required function of a mechanical system may therefore be defined by describing the interaction between the operator and the machine.

The function may then be modelled by describing the sequence of events which the operator and/or machine is to follow or by describing the actions of the operator, together with the sub-systems which make up the process.

DESCRIBING THE OPERATOR ACTIONS/SUB-SYSTEMS

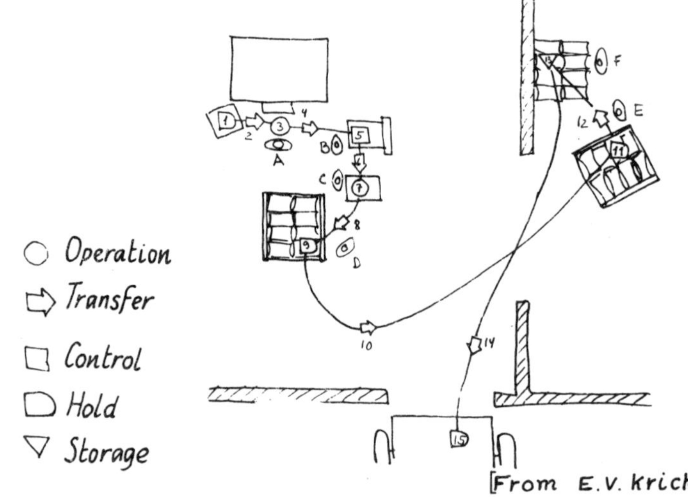

A sequence of operations requiring mechanisation. The operator's tasks and his equipment are shown. The code on the far left is used to identify the operations.

The sequence of hand operations in binding a book as a starting point to designing a mechanised binder.

DESCRIBING THE PROCESS ELEMENTS

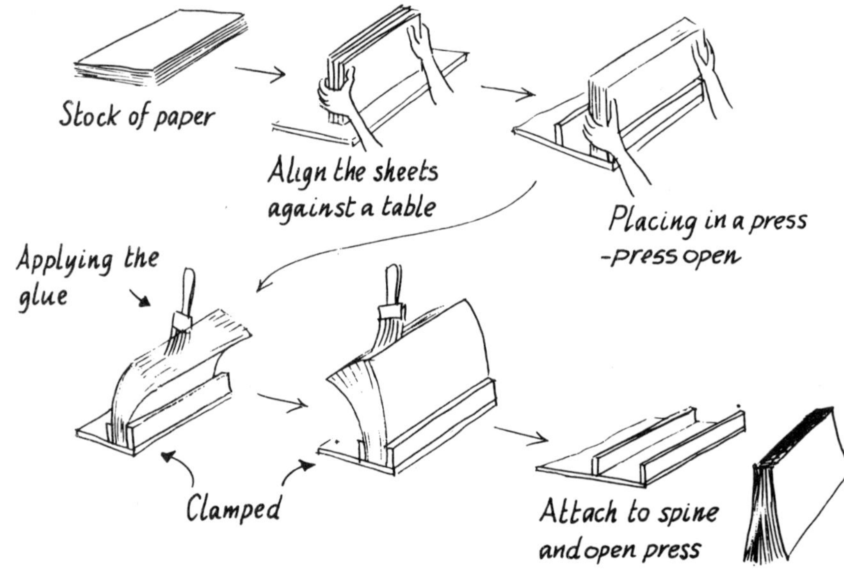

Modelling by Means of a Structure

In numerous design situations, it is necessary to model the function by first describing the structure; from this is implied the function.

Although the final objective of design synthesis may be to create a structure to achieve the desired function, it is bad practice to start by describing the function in terms of a specific structure since it may limit thinking, and better alternatives may be overlooked. Two extremes, one an abstract structure and one a complete structural drawing are shown below.

DESCRIPTION AS A SEQUENCE OF DRAWINGS

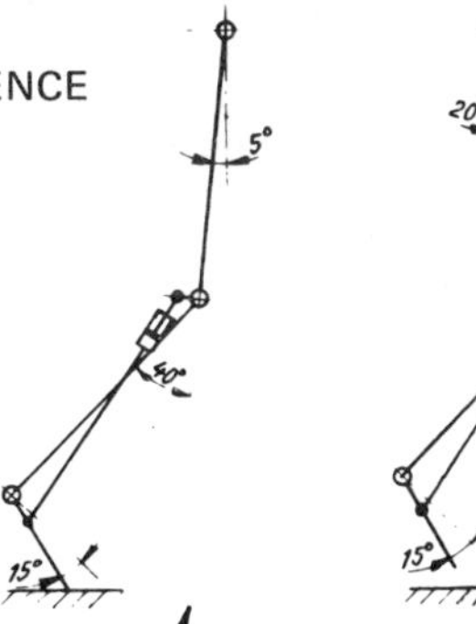

ABSTRACT DESCRIPTION

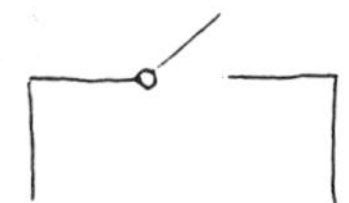

A simple abstract sketch is a good starting point for the design of the object, in this case a valve.

DESCRIPTION IMPLYING FORM

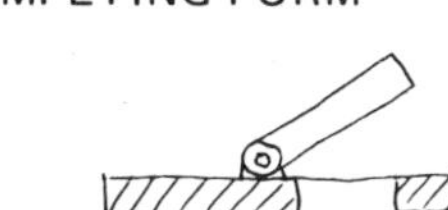

This sketch contains misleading details of form, and shows no more than the abstract sketch. The presence of the detail limits the designer's scope in developing form.

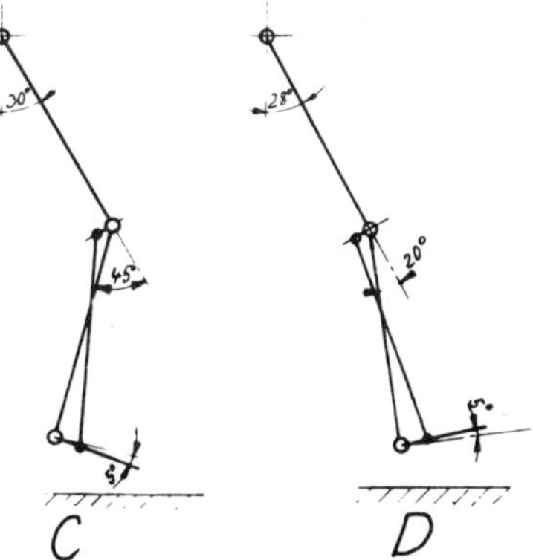

Line diagrams showing the function and the walking sequence of an artificial limb.

The final configurations of the valve; note that the sketch above contains no realistic details of the final form.

Modelling by Means of a Structure

PICTORIAL DESCRIPTION AS A SEQUENCE OF DRAWINGS

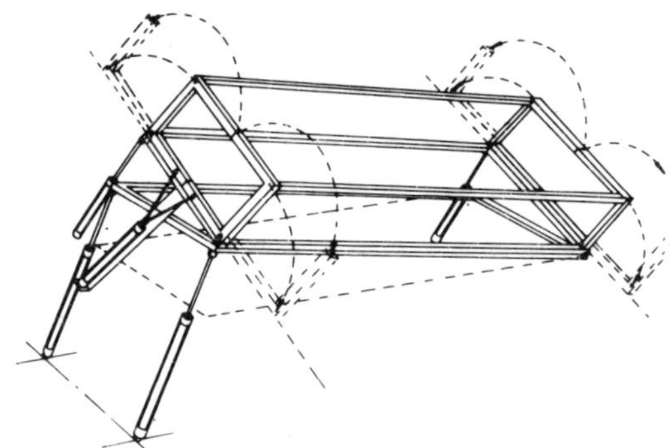

A pictorial drawing showing a framework in different positions. This drawing is the starting point for the design of a tipping system.

SYMBOLIC DESCRIPTION IN A SERIES OF DRAWINGS

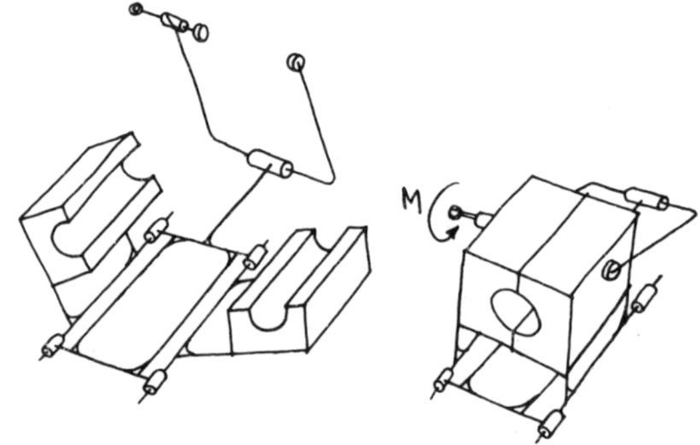

A symbolic drawing of a mechanism for the experimental production of sausage shaped eggs.

PICTORIAL DESCRIPTION

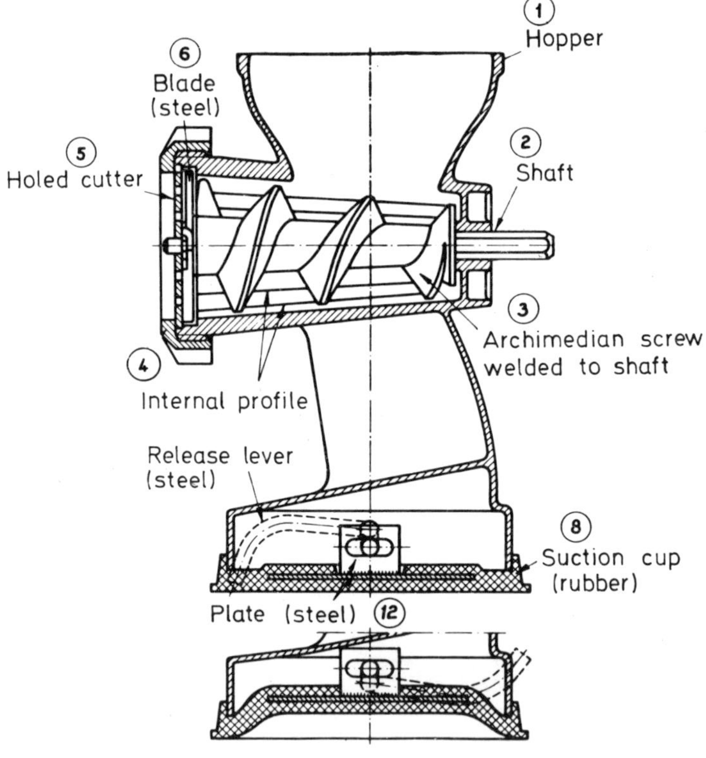

Section through a mechanical device which models a function.

STRUCTURE

Structure

The structure of a system consists of the elements in the system and their relationship.

Use

In the synthesis work, the designer is concerned with different aspects of structures. It can be seen from the sequence of properties:

required function → structure → form

that it is by establishing the structure that a principle is established according to which the required function is achieved. Therefore the transition from function to structure is central to the ideas stage. At the transition from structure to form, constructive shaping ensures that the principle can be realised.

MODELLING OF STRUCTURE

The structure for a product can be described at many levels; the following three will be discussed:

1 An abstract representation of system elements which in turn may be considered as sub-systems.

2 Symbolic representation of the elements of the system.

3 Visually representing the elements of the system.

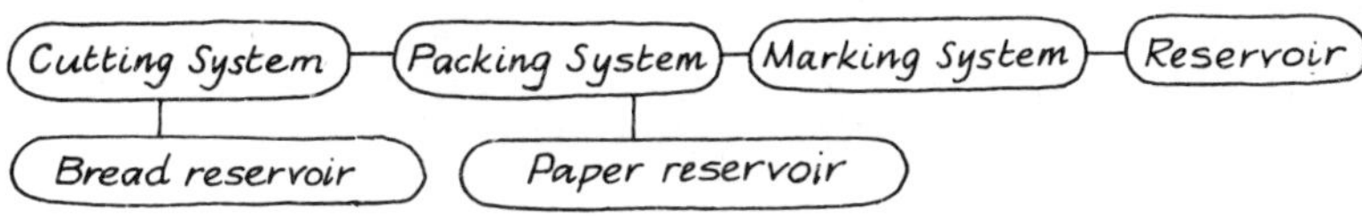

1 A block diagram of a mechanical system for cutting, packing and date marking bread.

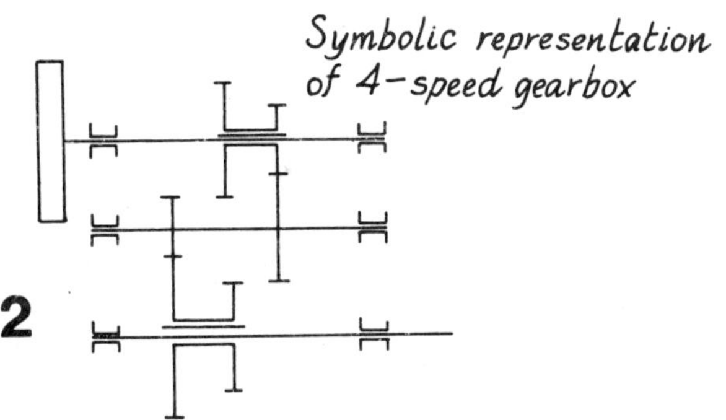

2

3 Sketch of flanged coupling

STRUCTURE 1.2

Abstract representation of sub-systems	Symbolic representation of elements	Visual representation of elements
1	2	3

Abstract Representation of Sub-Systems

The first step in structural modelling is usually the block diagram (see sheet 5.1). Each block represents a sub-system, comprising component parts, and the connecting lines represent physical relationships.

Example: Mechanical cutting, packing and date marking of bread.

Possibility 1

Cutting system — Packing system — Marking system — Reservoir

Bread reservoir

Paper reservoir

Possibility 2

Bread reservoir — Cutting system

Packing system — Reservoir

Paper reservoir — Marking system

Observations
Each diagram describes a system and, because the reader knows, or has an idea, what happens in a cutting or packing system, the general function is implied. The relationships (connecting lines) are, in this case, physical connections e.g. a conveyor belt or skid.

At the transition from structure to form it is still possible to experiment with the arrangement of parts. Here the problem is one of arranging the parts to best exploit the available space, bearing in mind that the parts require space.

Example: Vacuum cleaner

Observations
Although the full implications of the various structures have not been considered, it is apparent that the total form of the cleaner is dependent on the relative positions of the major sub-systems.

Symbolic Representation of Elements

When formulating ideas on possible solutions to a given problem, the designer requires a quick and convenient way of recording the structures which satisfy the functional requirements. These will be committed to paper to aid the designer in developing his ideas further and to communicate the ideas to others. Each engineering discipline has evolved symbolic notations. The following areas will be considered:

- Mechanical
- Hydraulic
- Pneumatic
- Electronic

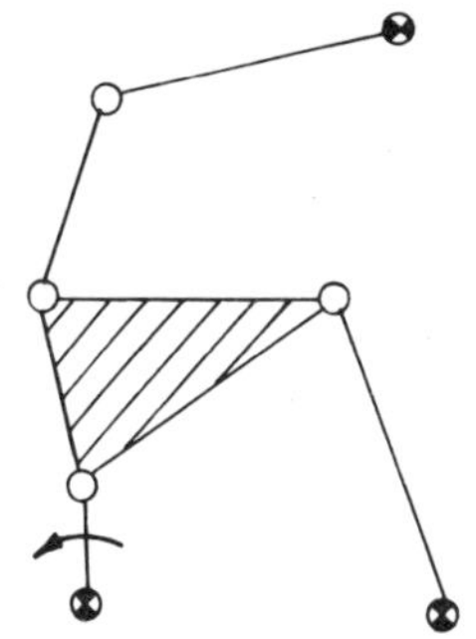

Planar six-bar linkage

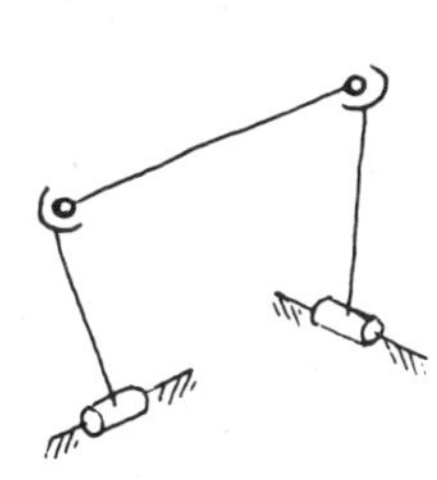

Spacial four-bar linkage

SYMBOLIC DESCRIPTION OF MECHANICAL SYSTEMS

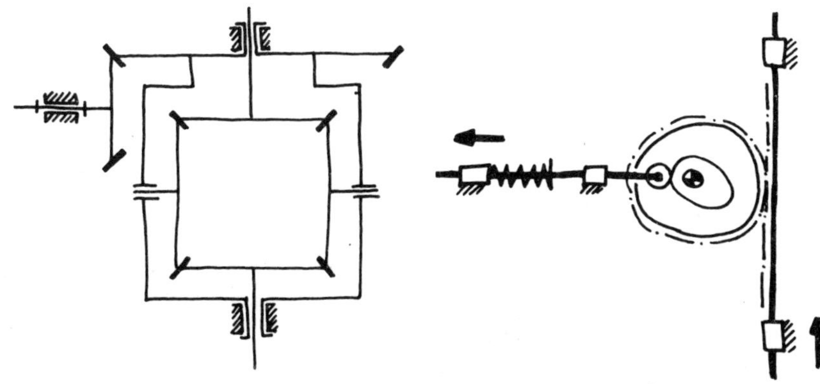

A differential

An arrangement for transmitting a linear movement through 90°

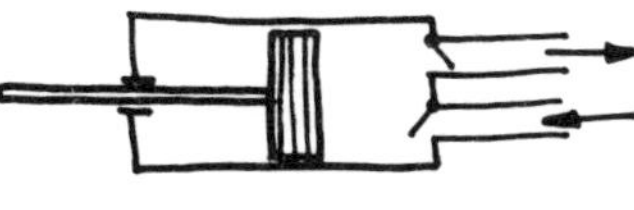

A piston pump

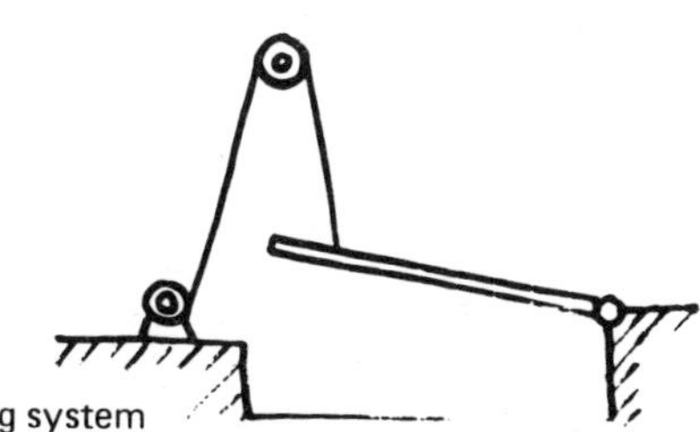

A tipping system

Observations

Each of these drawings shows a structure for the system and implies a principle of operation, i.e. the relationships between the elements are defined. Where possible the form of the element is omitted in order to emphasise the principle. Drawings of this type become the basis for detail design, i.e. the form design of the individual parts for the final product.

HYDRAULIC SYSTEMS

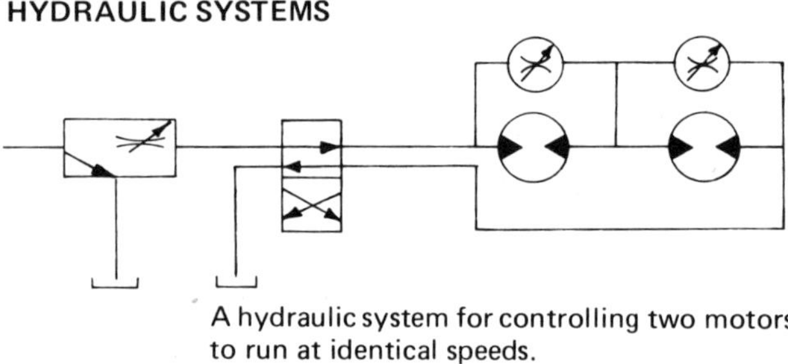

A hydraulic system for controlling two motors to run at identical speeds.

PNEUMATIC SYSTEMS

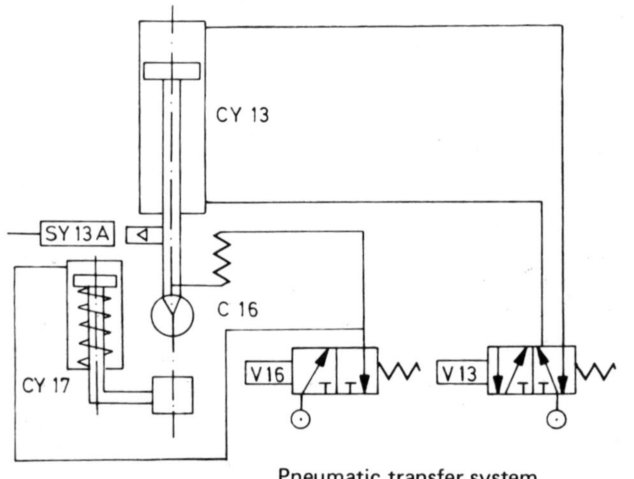

Pneumatic transfer system.

ELECTRICAL/ELECTRONIC SYSTEMS

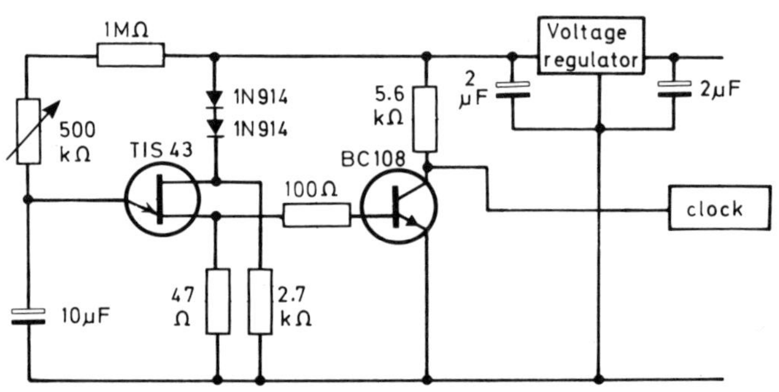

Electronic circuit for generating pulses for controlling an egg/sausage machine.

Observations

Each of these diagrams show the structure of the system by illustrating the relationships between the elements. The links in these relationships are provided by pipes and hoses and by wires and printed circuits.

Visual Representation of Elements

The structure of a form designed system can be modelled by a 'visual' drawing, showing the form of the elements. Often sectioning and other techniques have to be used to illustrate the system fully.

The following five categories will be considered:

direct visual representation.
section drawings.
layouts (see page 32).
arrangement or assembly drawings.
exploded views.
overlay drawings.

DIRECT VISUAL REPRESENTATION

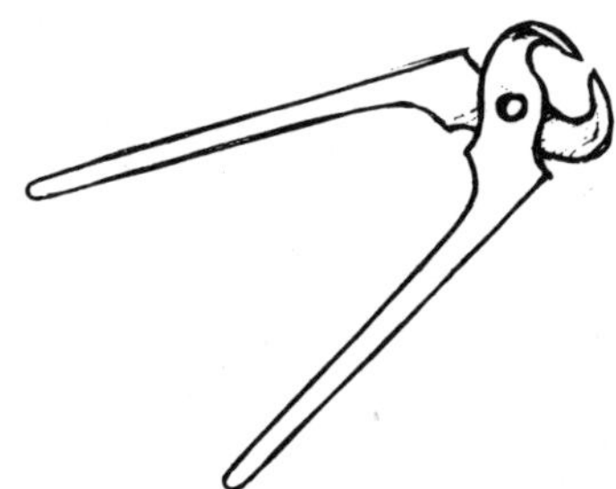

Observations:
The structure is apparent for simple systems or those which are predominently two dimensional.

SECTION DRAWING

Plane sections (orthographic projection)

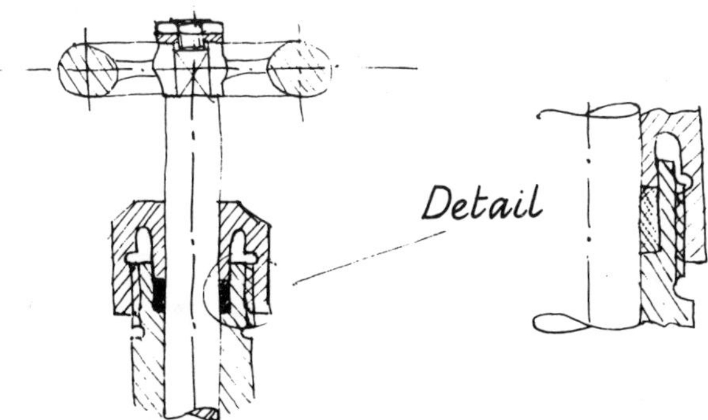

Observations:
Complete sections, should if possible be taken on a place about which there is some degree of symmetry; sections of this type are used in detail design, especially on layouts and assembly drawings.

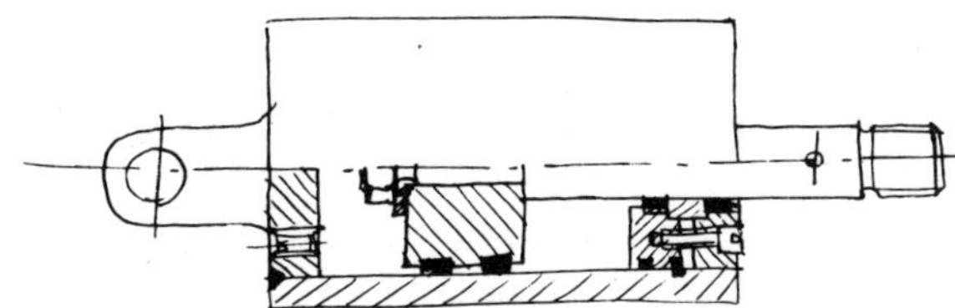

Observations:
Half section can be used where the object is symmetrical about a centre-line lying in the section plane. It enables internal and external views to be displayed simultaneously.

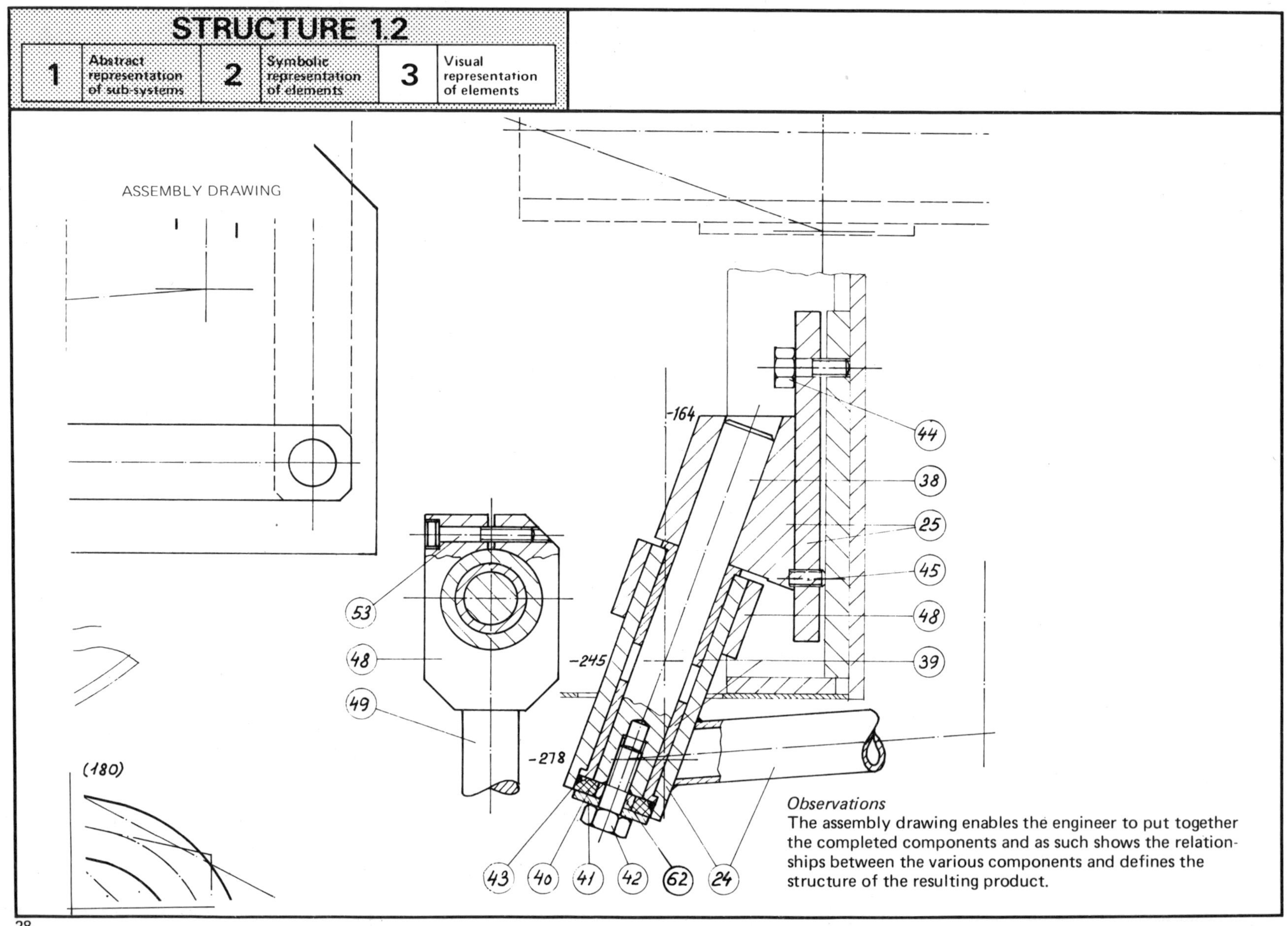

Observations

The assembly drawing enables the engineer to put together the completed components and as such shows the relationships between the various components and defines the structure of the resulting product.

INTERNAL VIEWS

Three dimensional sections

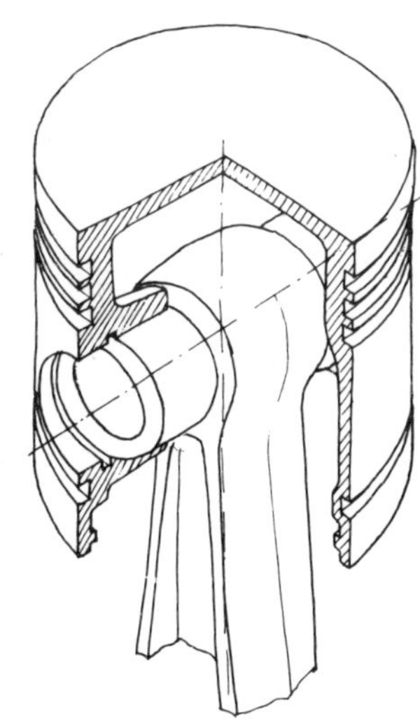

Quarter sections
Observations: Enables the inside to be seen on items which are symmetrical about a centreline.

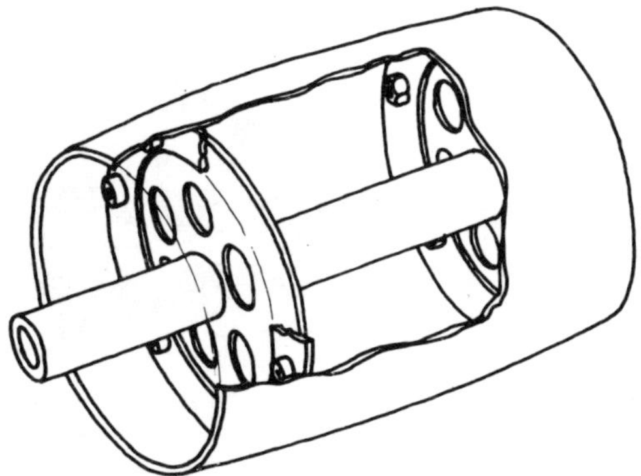

Cut-away drawings
Observations: The outer surface is cut away to show the interior. The cut 'edge' is shown.

OVERLAY DRAWINGS

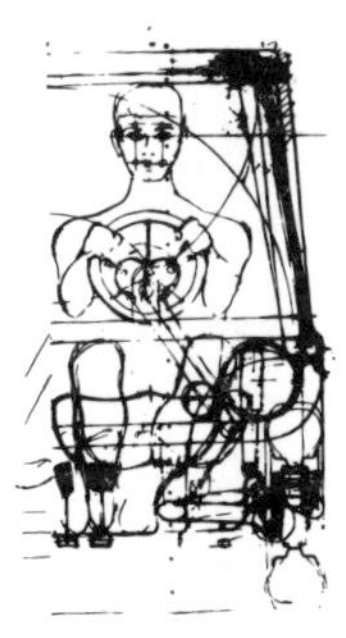

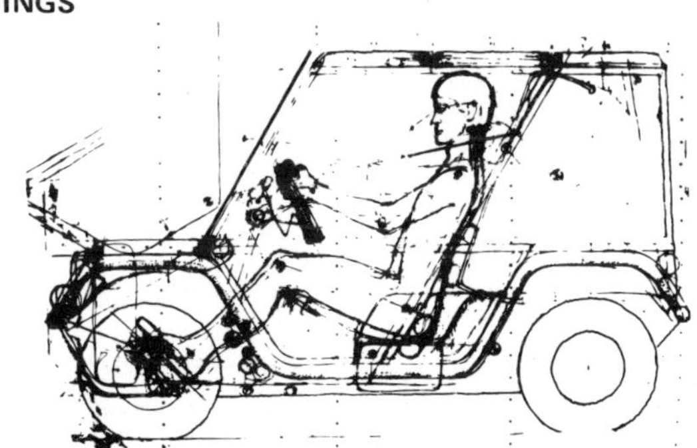

Observations: Drawings of the parts at different depths are superimposed to give a semi-transparent effect and show the relative positions of the inner components.

EXPLODED VIEWS

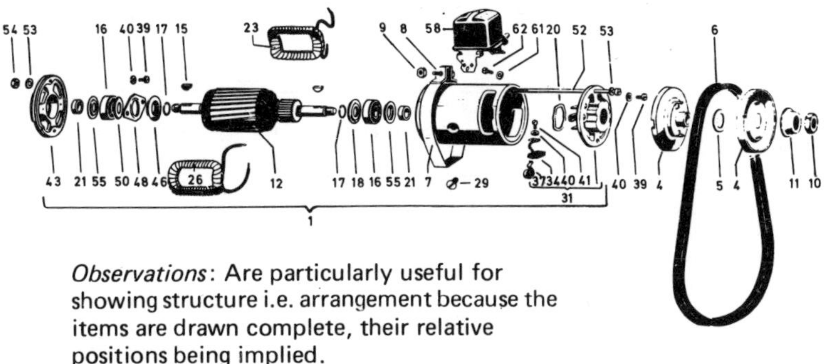

Observations: Are particularly useful for showing structure i.e. arrangement because the items are drawn complete, their relative positions being implied.

1.3	MODELLED PROPERTIES
FORM	

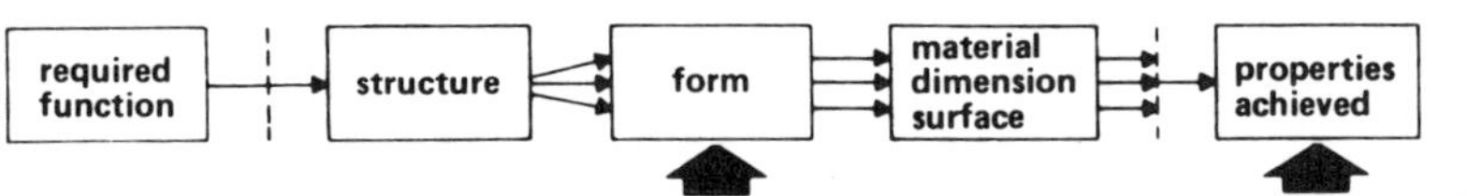

MODELLING OF FORM

Form is usually represented by a drawing projection i.e. a particular point on an object is transferred to a particular point on the paper. The projection may be formal, obeying certain geometrical rules, or it may be a direct visual image. With the latter, refer to procedure sheets 4.1.

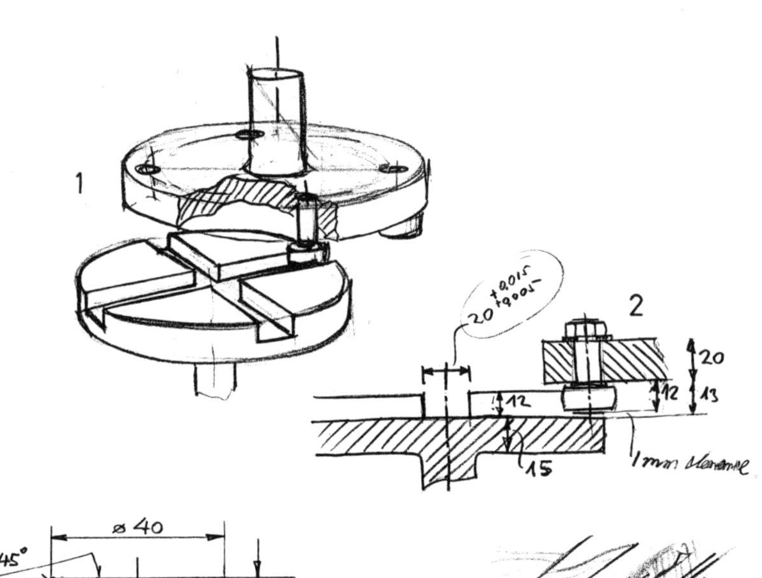

Uses

1 For DETAIL DESIGN, where the form is designed with the aid of sketches and layouts.

2 For COMMUNICATION WITH TECHNICAL DRAUGHTSMEN, who will prepare detail and assembly drawings.

3 For COMMUNICATION WITH PRODUCTION PLANNERS AND WORKSHOPS, using detail and assembly drawings.

4 For communicating the APPEARANCE of an object.

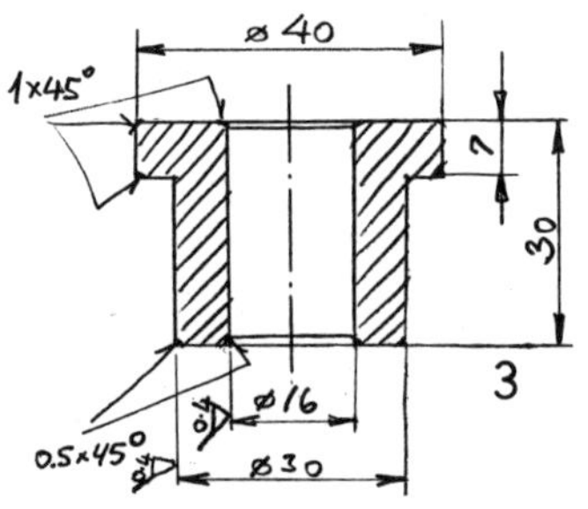

FORM 1.3

1	Detail design	2	Communication with technical draughtsmen	3	Communication with production planners and workshops	4	Appearance

Form-Detail Design

The form of an item is decided by detail design. The starting point is the structure and the objective is an integrated form together with specified materials and finishes. The process: structure to form, requires very intense self communication by the designer and here FREEHAND SKETCHES and LAYOUTS are invaluable modelling aids.

FREEHAND SKETCHES

Example: Bailing pump for bailing rain water out of small unattended boats at a mooring.

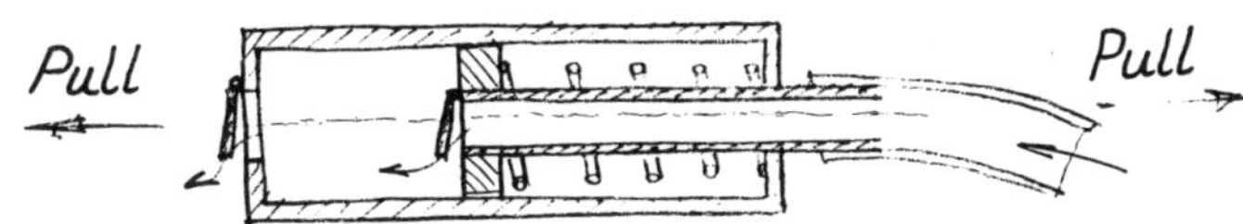

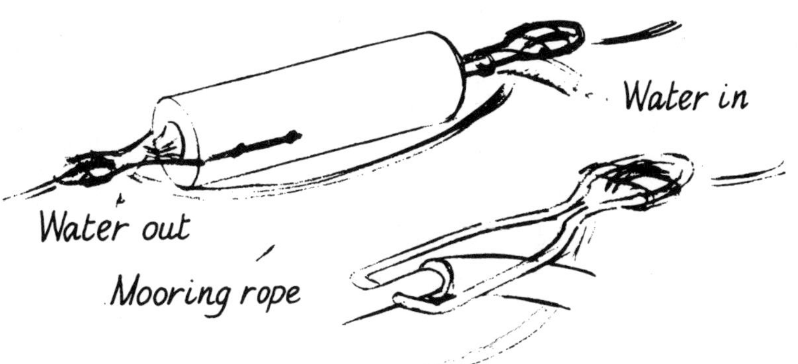

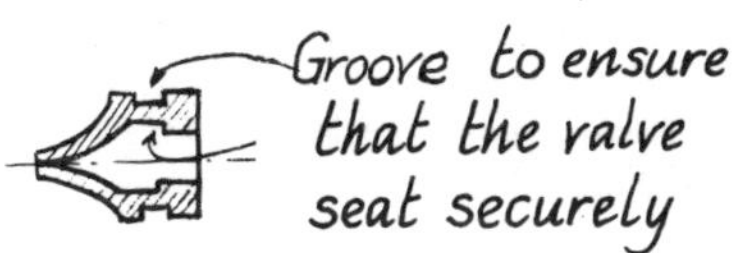

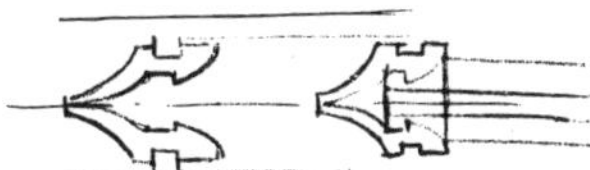

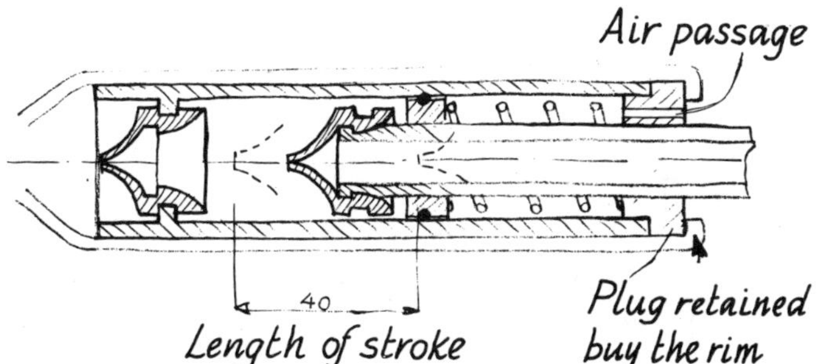

Observations
Sketching is an all embracing term, as every model drawn without the aid of instruments becomes a freehand sketch. Sketches are informal; only necessary details are drawn and the type of projection is chosen according to need. Sketching in this context, should be speedily implemented to avoid breaking the line of thought and the temptation to expand in detail resisted.

LAYOUT

Observations

In the final phase of design layouts are used. The form of the individual elements is decided progressively using a number of steadily improved layouts. Drawing layouts to scale is therefore important.

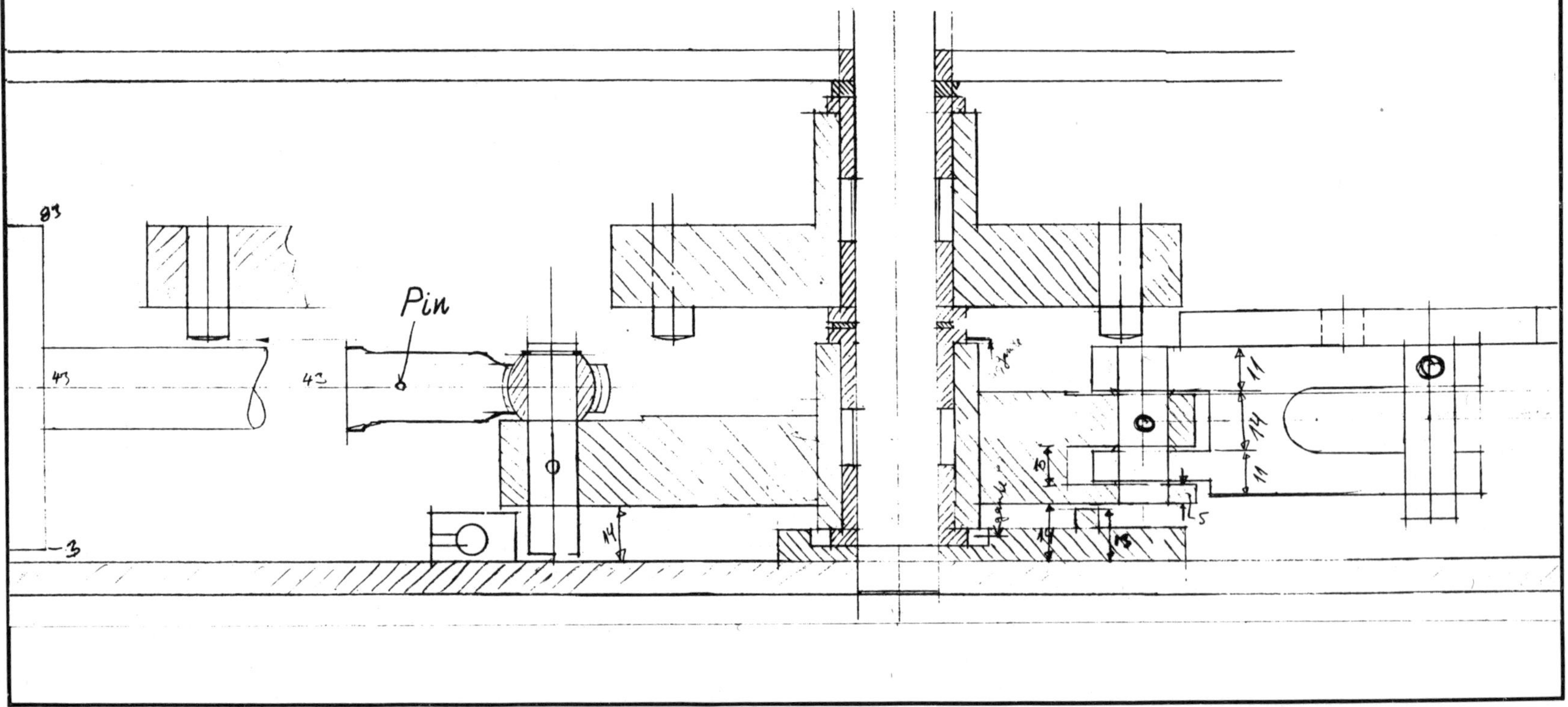

Form-Communication with Technical Draughtsmen

A technical draughtsman who is to prepare detail and assembly drawings, must have a basis on which to build. This is normally provided as sketches and layouts supplemented by necessary data on materials, dimensions and surface qualities.

THREE DIMENSIONAL SCALE SKETCHES

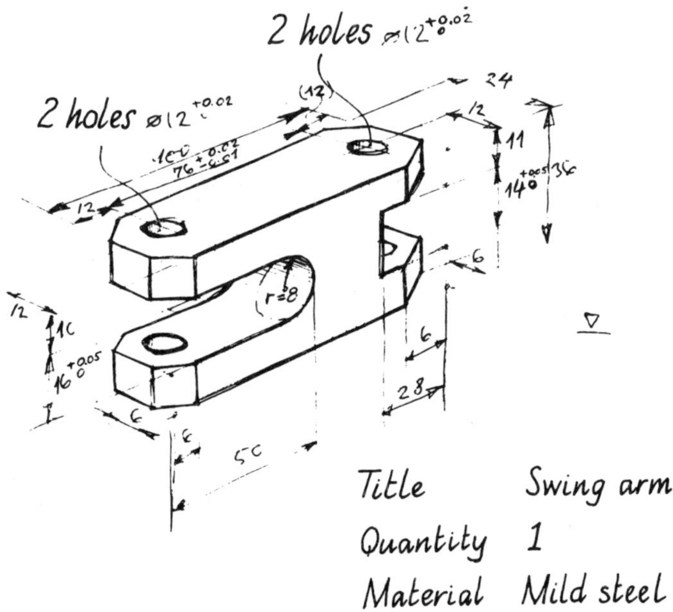

Observations
This perspective sketch gives all the information about the item and may be used by a draughtsman to prepare formal detail drawings or used as a detail drawing in its own right.

SKETCH IN FIRST ANGLE PROJECTION

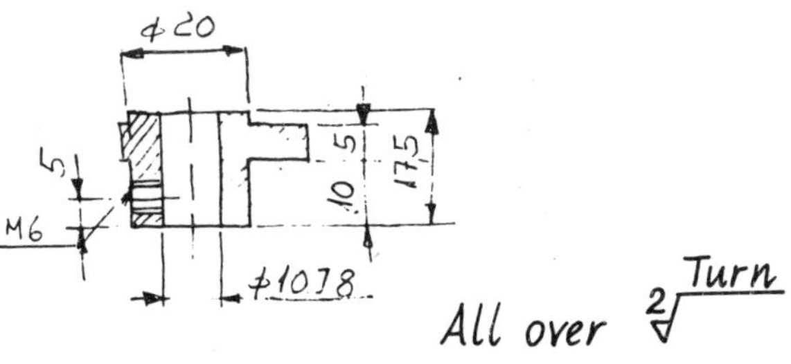

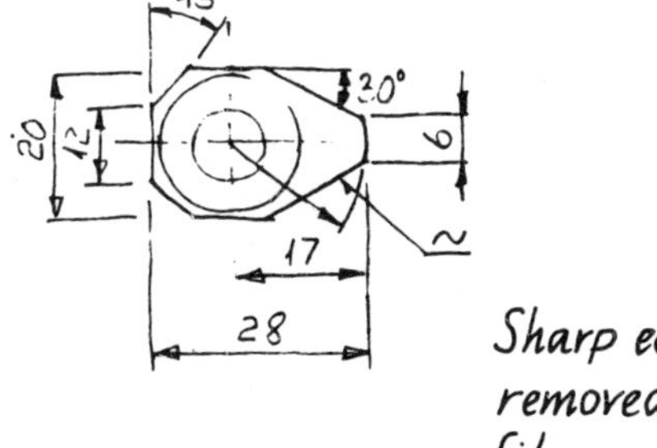

Observations
A rough sketch is often produced in advance of a formal detail drawing. How comprehensive the dimensions are depends on circumstances. Further measurements can often be taken from an accompanying layout or from existing detail drawings of mating objects.

Form - Communication with Production Planners and Workshop Staff

An item to be manufactured in a workshop, is normally specified by a detail drawing. Sketches may be used in some circumstances, either as 'freehand – detail drawings' or as 'pictorial sketches' with adequate dimensions and other necessary information. Complex systems are described by a number of detail drawings together with assembly drawings and parts lists.

Planners and workshop staff generally need the same information. The planner should be able to see, from the drawing, which operations are to be performed in the workshop.

DETAIL DRAWING

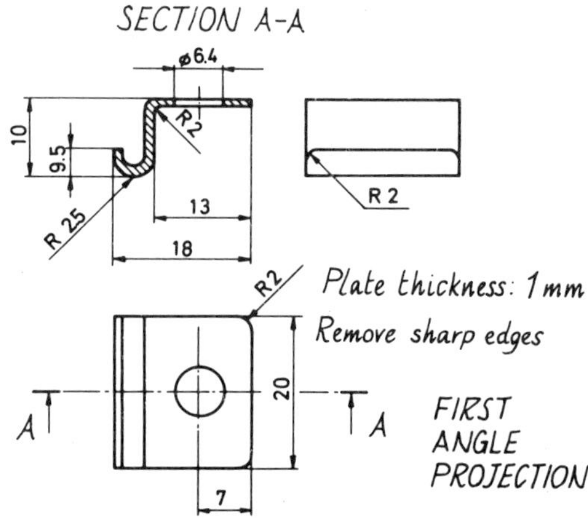

Observations

Detail drawings are normally drawn according to National Standards, in the UK, this is BS 308 (ISO/R128, ISO/R129, ISO/R406, ISO/R1302 & ISO/R1101). The rules have evolved from tradition and experience. Drawings may be produced in pencil or drawing ink on a variety of semi-transparent media depending on the degree of durability required and on national preferences.

FREEHAND DETAIL DRAWING

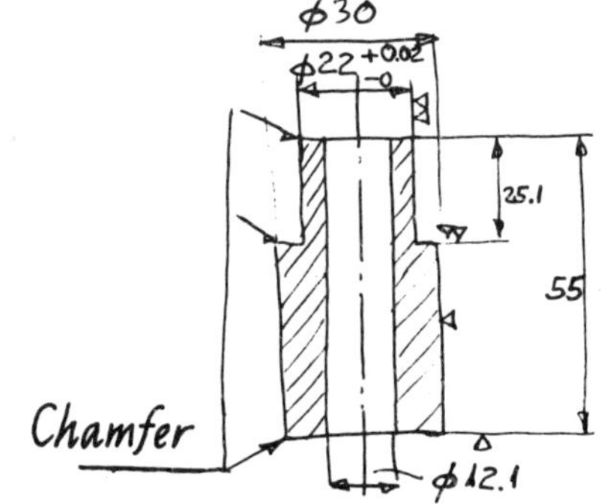

Description Bush

Quantity 2

Material Brass

Observations

Freehand detail drawings are normally used where only a few items are required, e.g. one-off manufacture, prototype construction.

DIMENSIONED PICTORIAL SKETCHES

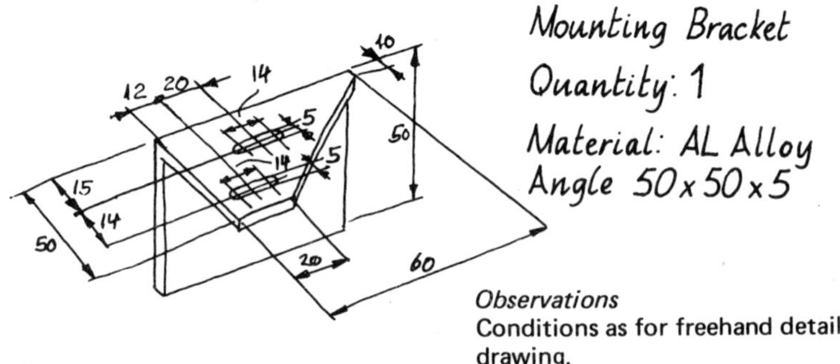

Mounting Bracket

Quantity: 1

Material: AL Alloy Angle 50 x 50 x 5

Observations

Conditions as for freehand detail drawing.

Form - Appearance

Appearance is dependent on form, colour and surface texture.

Form is fundamental in the graphic modelling of appearance, whilst colour and texture can normally only be specified.

The graphic model should as far as possible give the same visual appearance as the 'original'. Therefore perspective or perspective like projections are used.

Form may be emphasised by shading and in certain areas, by drawing contours, or a net of squares, etc. See work sheet on page 78.

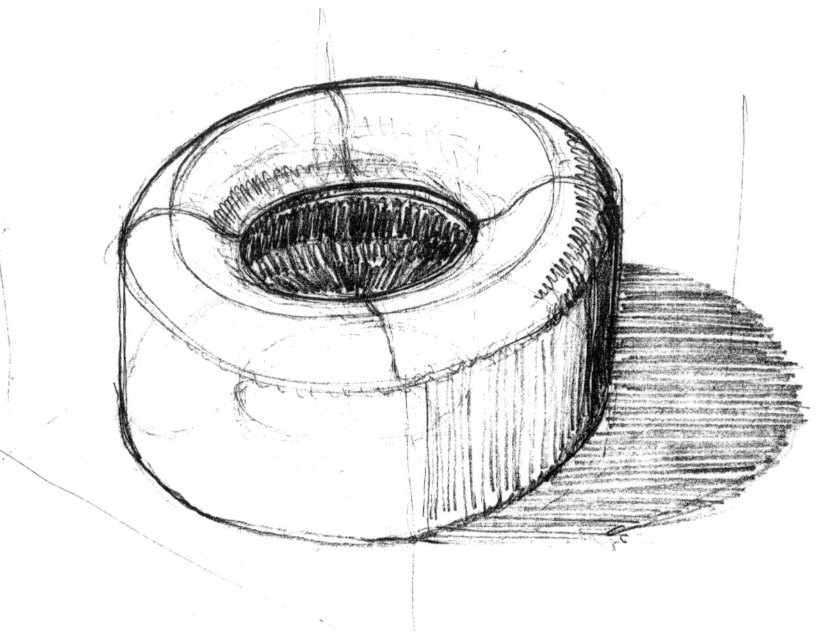

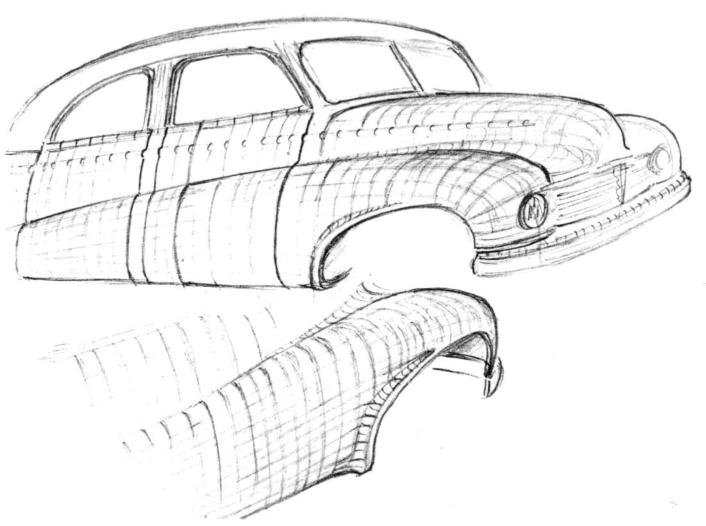

Observations
Note the effect of shading. Sometimes a few lines show the form clearly.

Observations
This technique is used effectively on double curvature. It can, if necessary, be supplemented by shading. See work sheet 5.12.

MATERIAL

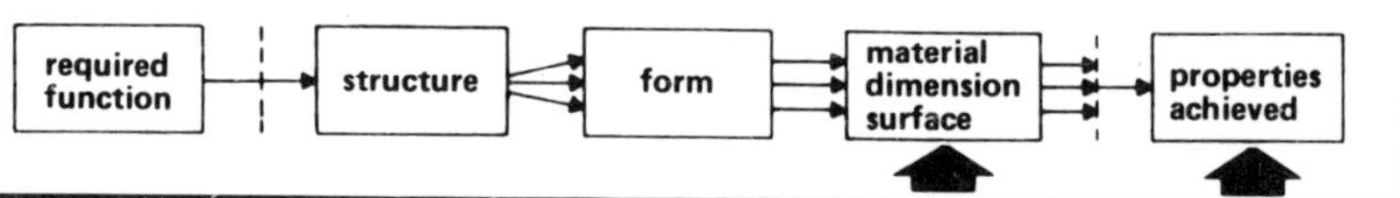

Material

For every element in a mechanical system, a material must be chosen, so that the item can be produced and function as intended.

Use

One may have the need to communicate a material specification to a designer or technical draughtsman. It is always necessary in communications with the workshop.

CODES USED IN SECTION DRAWINGS

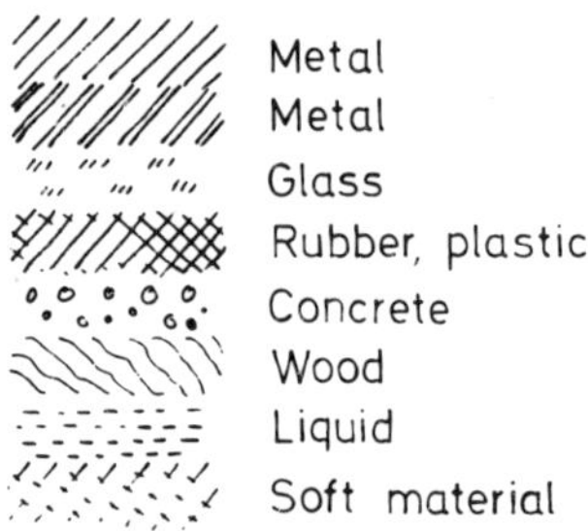

Examples of symbols that might be used in section drawings (standard symbols may be recommended by specific organisations, the most important consideration is that the codes must be meaningful, and that they must be consistent within a given drawing or project).

MODELLING BY A WRITTEN STATEMENT

Materials are usually described by a written statement and the following are normally required to define the material correctly: common type name, e.g. mild steel; the material specification, usually complying to some industrial standard; the form taken by the raw material; the treatment process to which it must be subjected. The specifications vary between countries and between indistries.

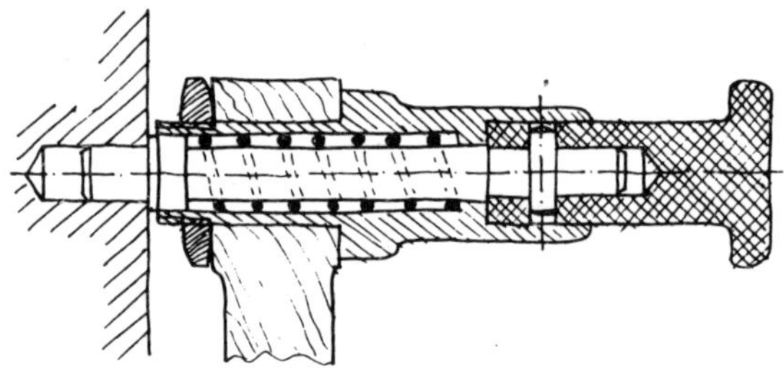

Assembly sketch using sectioning codes.

DIMENSION

Any item must be dimensioned in such a way that it can be manufactured and also function as intended.

Use

It is important to distinguish between the functional dimensions and dimensions which can be chosen more freely, i.e. only limited by production demands, appearance and economy. The need to specify more than the functional dimensions depends on the situation, but in practice the designer is normally expected to furnish most of these dimensions. Technical draughtsmen and workshop staff must have a starting point for their activity, e.g. partially dimensioned sketches and complete working drawings respectively.

MODELLING DIMENSIONS

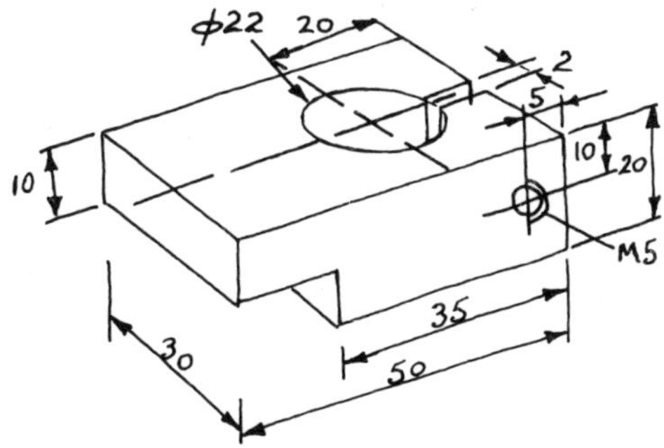

A dimensioned sketch of an item which the designer requires as a formal detail drawing.

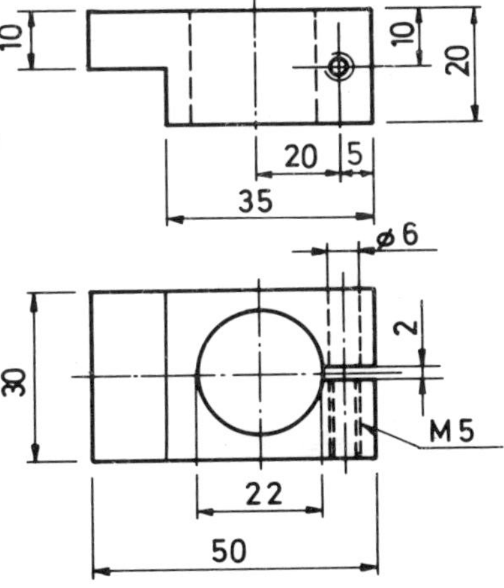

The same item drawn according to British Standards.

A toleranced metric drawing employing standard tolerancing.

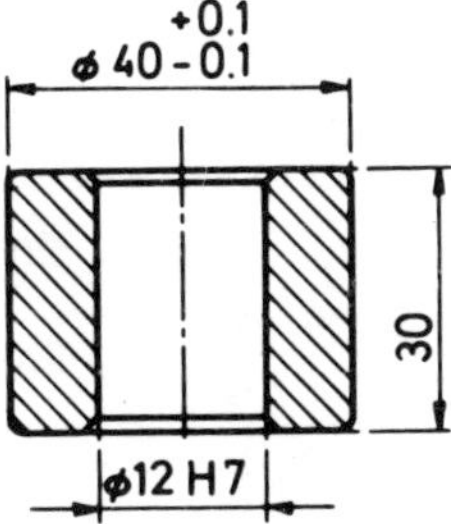

It is a common practice for designers to work to systems of general tolerances. Here tolerances are preprinted as a key on the drawing sheet. The key indicates the tolerance applicable for a given preciseness of dimensioning i.e. the tolerance is related to the number of decimal places quoted on a dimension.

1.6 MODELLED PROPERTIES

SURFACE

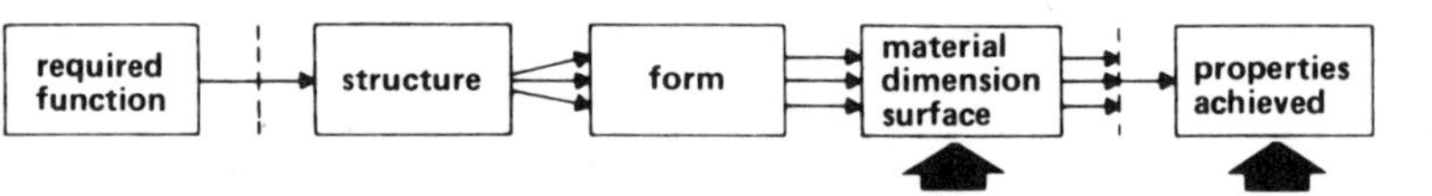

Definition

The surface is specified in such a way that the item can be manufactured and function as intended. When machining processes are involved this parameter is normally referred to as 'surface finish' or 'finish'.

Use

It is important to emphasise that surface may be decided by function, but may also be chosen freely, only limited by the demands of production and economy.

The reason for specifying the surface is to convey the designer's ideas for the quality of the surface required. This is achieved by describing the finishing process.

MODELLING SURFACE

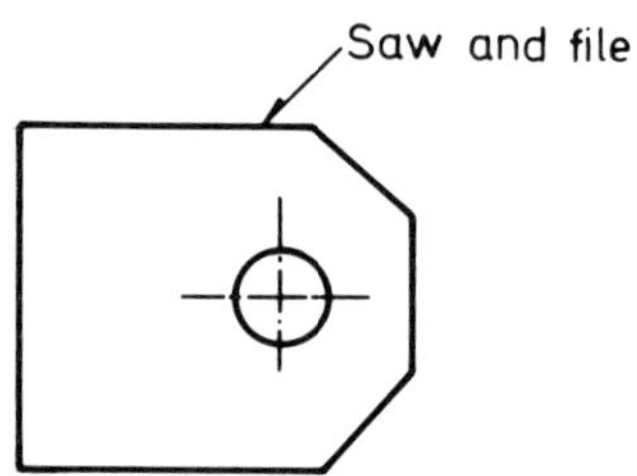

50

Tinned

Information about tinning the end of a spigot.

Standardised specification for surface finish.

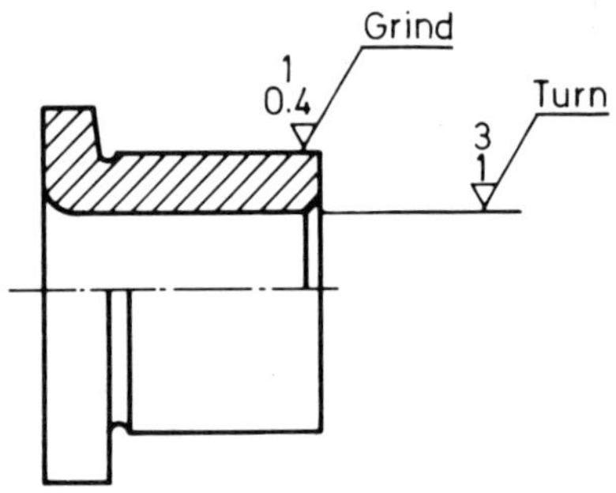

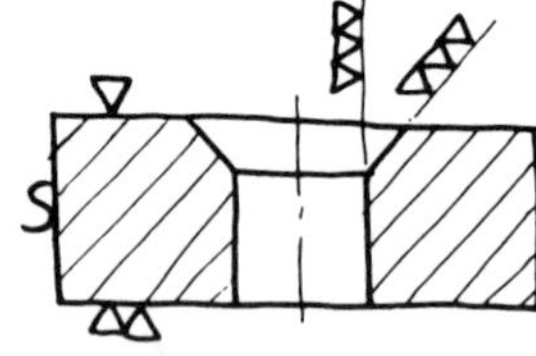

An informal way of indicating the quality of surface finish.

2. THE RECEIVER

2.0 SUMMARY

EXAMPLES

SUMMARY

When a drawing is produced, it is necessary to know the user. Only when the 'receiver' is known, can the drawing codes be decided. In other words: the receiver sets the criteria by which the drawing must be judged. If the receiver understands the codes, and if the drawing is easily read and in use, then the drawing fulfils its purpose.

There are two categories of receiver: the designer himself and others. A more detailed division is given below.

Receiver: the designer
Self communication through the preparation of a sketch or drawing, is an important factor in design. The drawing supplements the brain during creative work, when ideas have to be formulated, made concrete, investigated and appraised. The draughting process stimulates the imagination because it is impossible to visualise all the possibilities before commiting pencil to paper. The drawing reveals obscurities, raising questions and giving opportunities for new ideas. Creative work is considerably strengthened by this form of self communication. Drawings produced in this way, may show considerable interdependence between sketches and notes.

Speed is important in sketching and drawing, if it takes too long, ideas may die or disappear before the drawing is finished. The skill of speedy sketching and drawing is fundamental to the design engineer. The technique is to concentrate on the main outlines excluding unnecessary and distracting details.

Receiver: Another designer
This may be a colleague with whom the designer is developing the project, or a designer taking over part of the job. In both cases it is important that conversation and drawing become complimentary.

Receiver: Draughtsman
A draughtsman develops the work in greater detail to produce detail drawings. He therefore starts from information in the form of sketches and layouts, supplemented by verbal comment.

Receiver: Workshop and production planners
When the designer communicates with workshop staff and production planners, it is with a view to manufacturing the item. The information must be clear and precise, so that the drawing is self-explanatory. Standardised drawing types are normally used, i.e. detail and assembly drawings. Other types are used in special situations e.g. the use of three dimensional sketches for speed or because of limited quantities.

Receiver: Directors, client, professional group
This group needs a given amount of concise information. Additionally, the project or ideas must be 'sold' to the recipients. This is demanding on the drawings, which often take the form of technical illustrations.

Receiver: The public
The media for communication are articles and books. These are normally one way unsupplemented communications. Such communication is particularly demanding on drawing skill because of the demand on easy reading and the demands from the reproduction technique.

Example

A component is shown for three different recipients: the designer, draughtsman and workshop staff. The first is a layout, drawn using a draughting machine. The second is a three dimensional dimensioned freehand sketch. Finally a detail drawing for workshop use, drawn using a draughting machine. It is common practice to produce formal drawings in pencil, but many European countries have adopted the practice of drawing in ink.

RECEIVER: THE DESIGNER

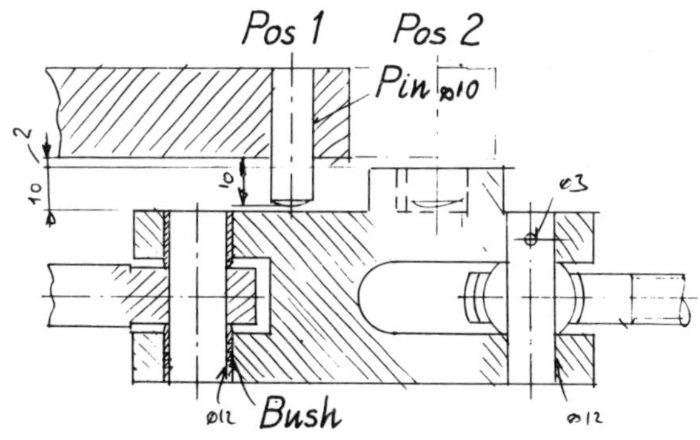

RECEIVER: TECHNICAL DRAUGHTSMAN

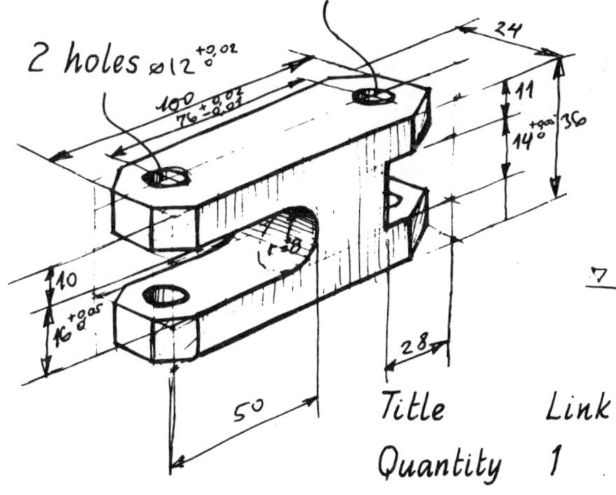

RECEIVER: WORKSHOP

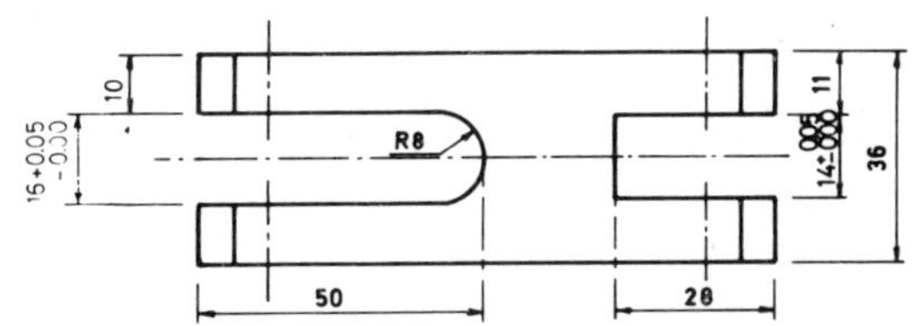

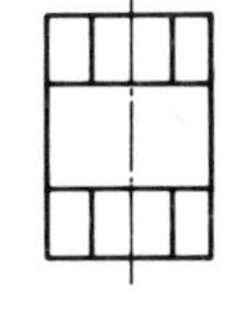

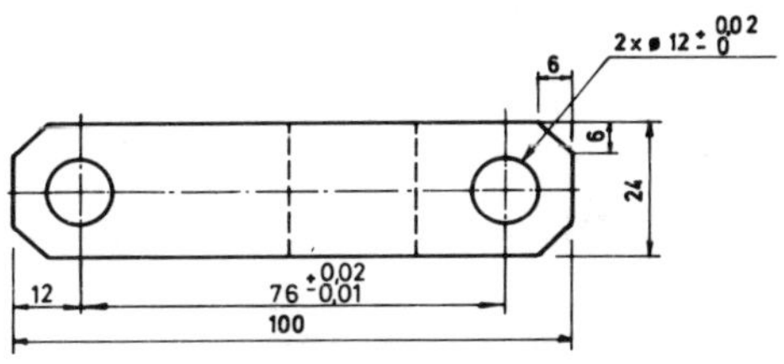

THE RECEIVER 2.0

Example

Here plate shears are drawn for three different recipients: the designer himself, a technical group and client. The first two drawings use mechanical symbols, whilst a projection has been used for the client. The respective techniques are: freehand sketching, bold freehand drawing and drawing with instruments.

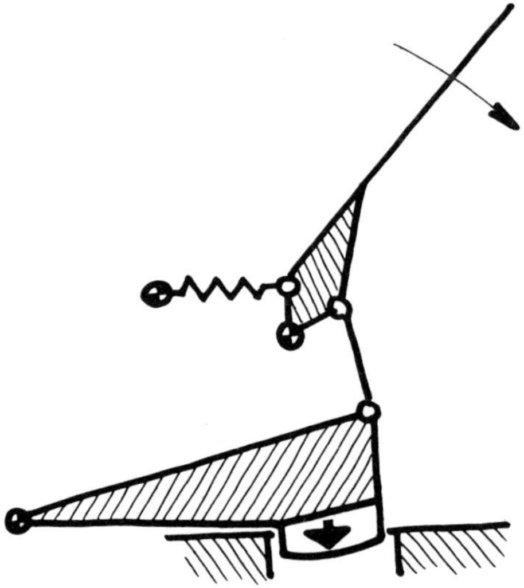

RECEIVER: TECHNICAL GROUP

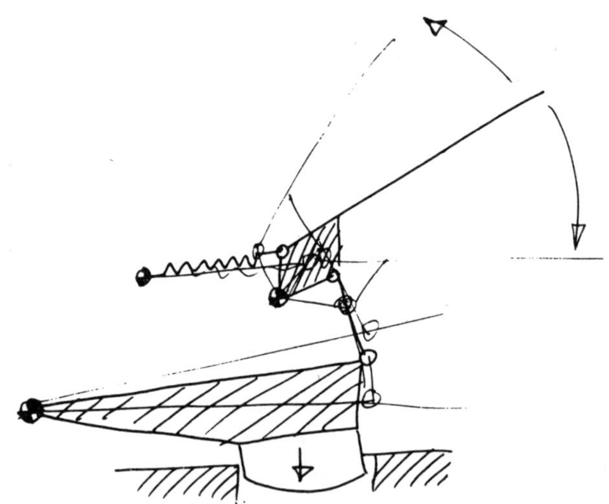

RECEIVER: THE DESIGNER HIMSELF

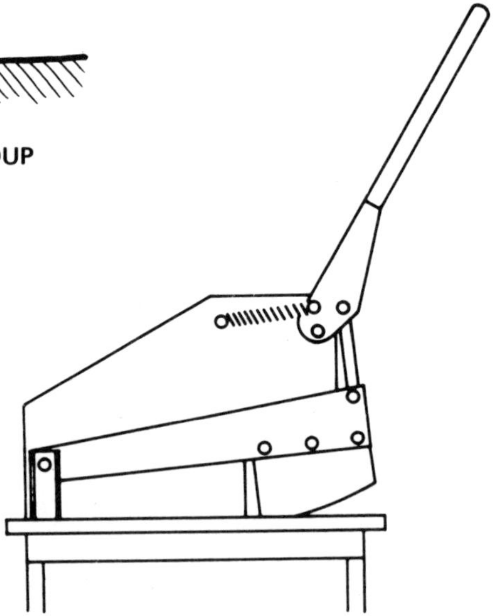

RECEIVER: CLIENT

3. THE DRAWING CODE

3 0 SUMMARY

3.1 CO-ORDINATES

3 2 SYMBOLS

3.3 TYPES OF PROJECTION

3.0	THE DRAWING CODE

SUMMARY

Codes are sets of rules according to which the information on a graphical model is presented. Coding exists where use is made of: e.g. pneumatic symbols, thread notations or where three-dimensional objects are represented as isometric drawings.

Graphic models are primarily for communication. For the drawing to be interpreted correctly, the code used in formulating the information should be known to the recipient of the message i.e. the receiver should know pneumatic symbols, thread notation and isometric representation.

It is important that in a given situation the code is chosen correctly, so that the receiver understands the drawing. The code is chosen for simplicity (mechanical symbols), precise definition (mechanical drawing) or conveying information (three-dimensional representation) or some other properties.

CO-ORDINATES

Translation into co-ordinates (encoding) and graphical representation of these co-ordinates are important aids to demonstrating the interdependence of the various parameters.

There may be many different sorts of parameters, such as the speed of a sewing machine needle as a function of its position, the amount of sagging in a load-bearing girder at each point in the girder, the position of a machine part in a machine system, etc.

In turn the visual image provides the basis for vector algebra, which is a mathematical way of representing what can be complex graphical situations. The basic concept of representing a vector by components parallel to the Cartesian axes is shown below.

Vectors having magnitudes of unity and denoted $\underline{i}, \underline{j}, \underline{k}$ represent the directions of the axes X, Y, Z. A vector is then represented as

$\underline{R} = a\underline{i} + b\underline{j} + c\underline{k}$ where a, b and c are the magnitudes of the vector components.

This form of vector representation makes possible standard algebraic vector manipulations and simplifies the description of certain graphical forms.

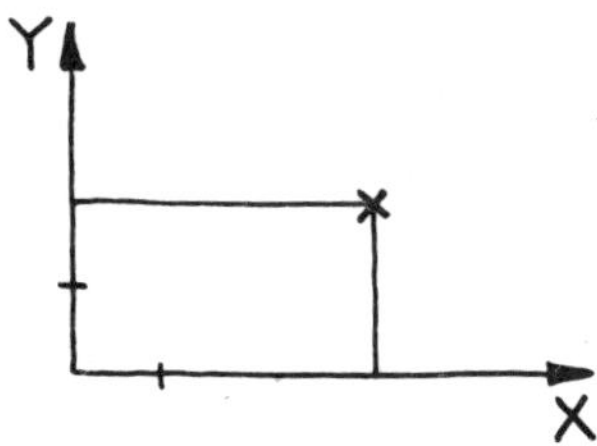

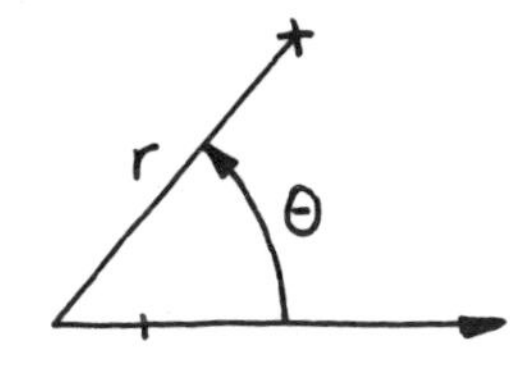

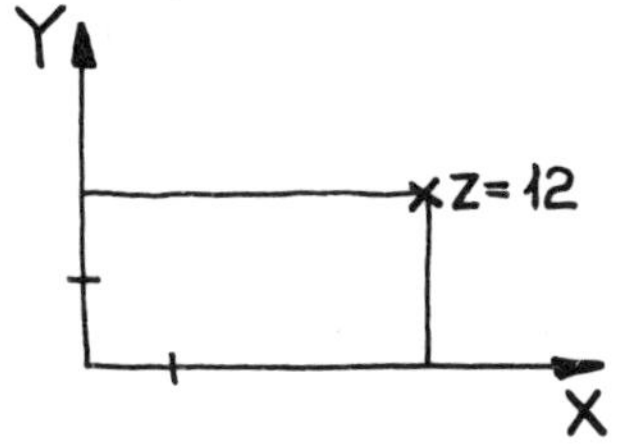

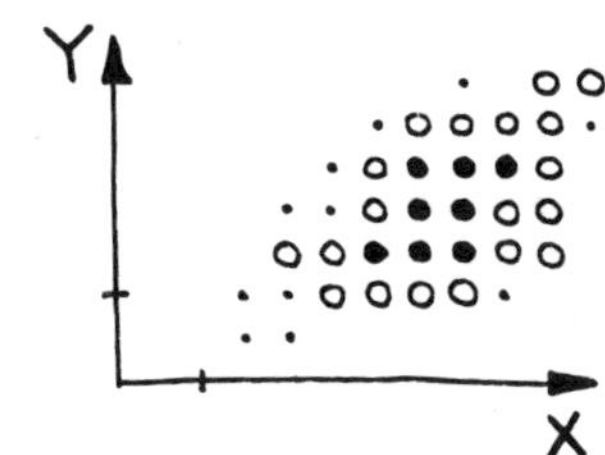

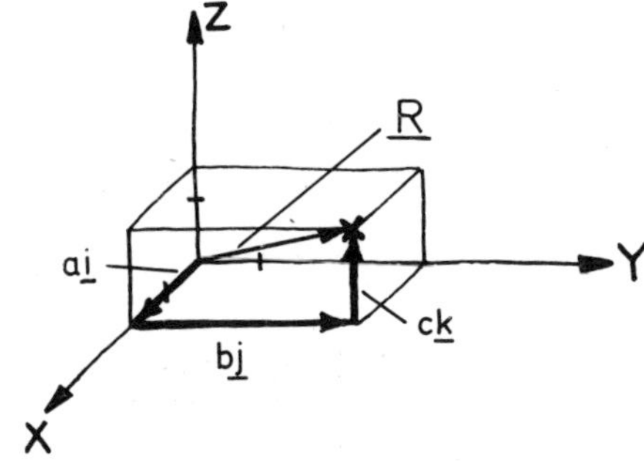

3.2 THE DRAWING CODE

SYMBOLS

A symbol is a notation or simplified picture with a specific meaning. It contains unmistakable coded information.

The characteristic of a symbol is that, when used on a drawing, it gives clear and distinct information in a few quick strokes.

Symbolic representation is very diverse, the most relevant groups for mechanical engineers being:

- Mechanical symbols
- Hydraulic/pneumatic symbols
- Electrical/electronic symbols
- System symbols
- Mechanical drawing conventions, e.g. notations for: threads, diameters, centre lines, roughness.

Apart from the above a designer may devise his own additional symbols.

Symbols are used in the following types of drawing:

- Drawings of principle
- Detail and assembly drawings
- Diagrams

SYMBOLS 3.2

1	Mechanical symbols	2	Hydraulic and pneumatic symbols	3	Electrical/ electronics symbols	4	System symbols	5	Drafting conventions

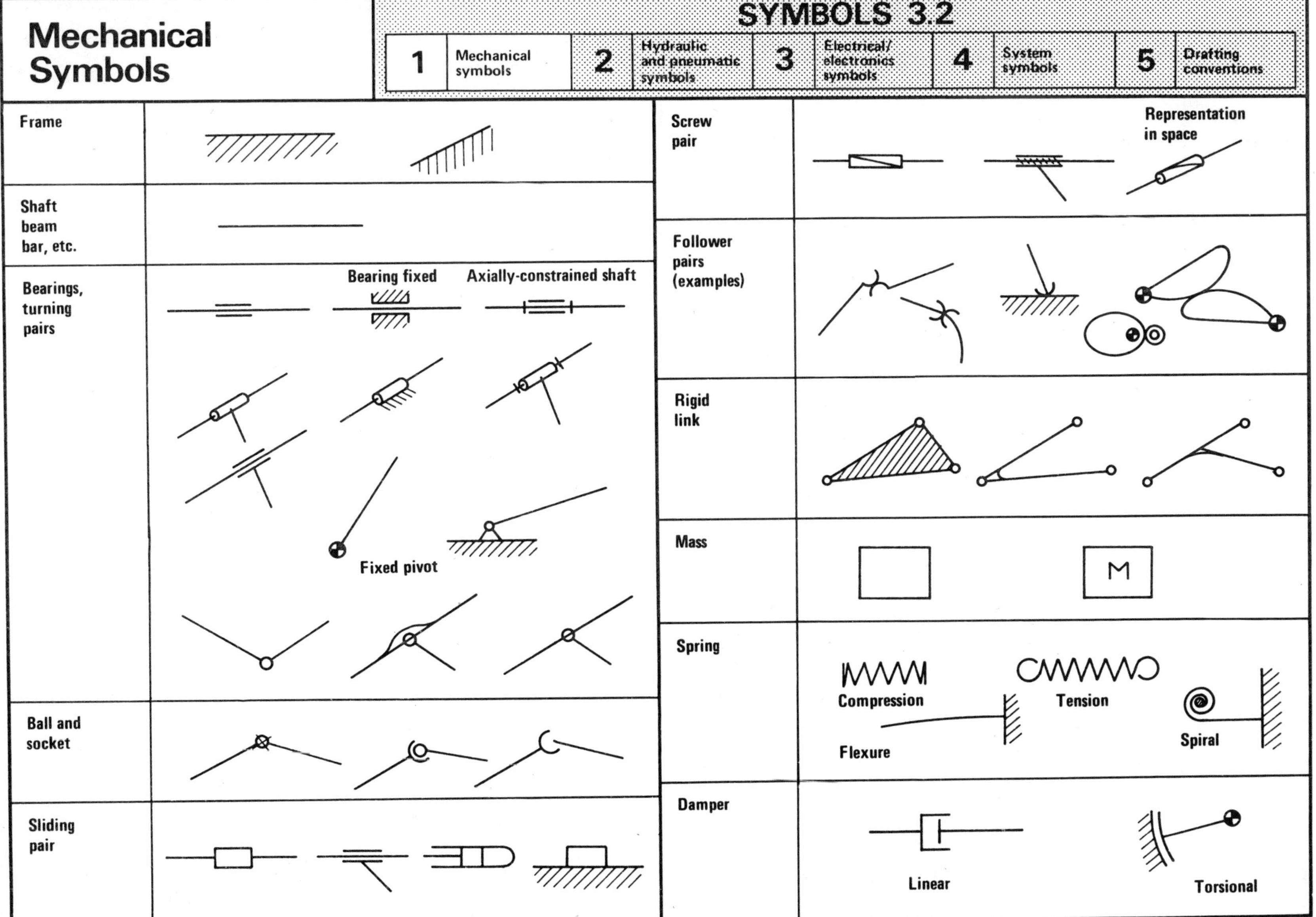
Mechanical Symbols
SYMBOLS 3.2
1 Mechanical symbols
2 Hydraulic and pneumatic symbols
3 Electrical/ electronics symbols
4 System symbols
5 Drafting conventions
Frame
Shaft beam bar, etc.
Bearings, turning pairs
Bearing fixed
Axially-constrained shaft
Fixed pivot
Ball and socket
Sliding pair
Screw pair
Representation in space
Follower pairs (examples)
Rigid link
Mass
M
Spring
Compression
Tension
Flexure
Spiral
Damper
Linear
Torsional

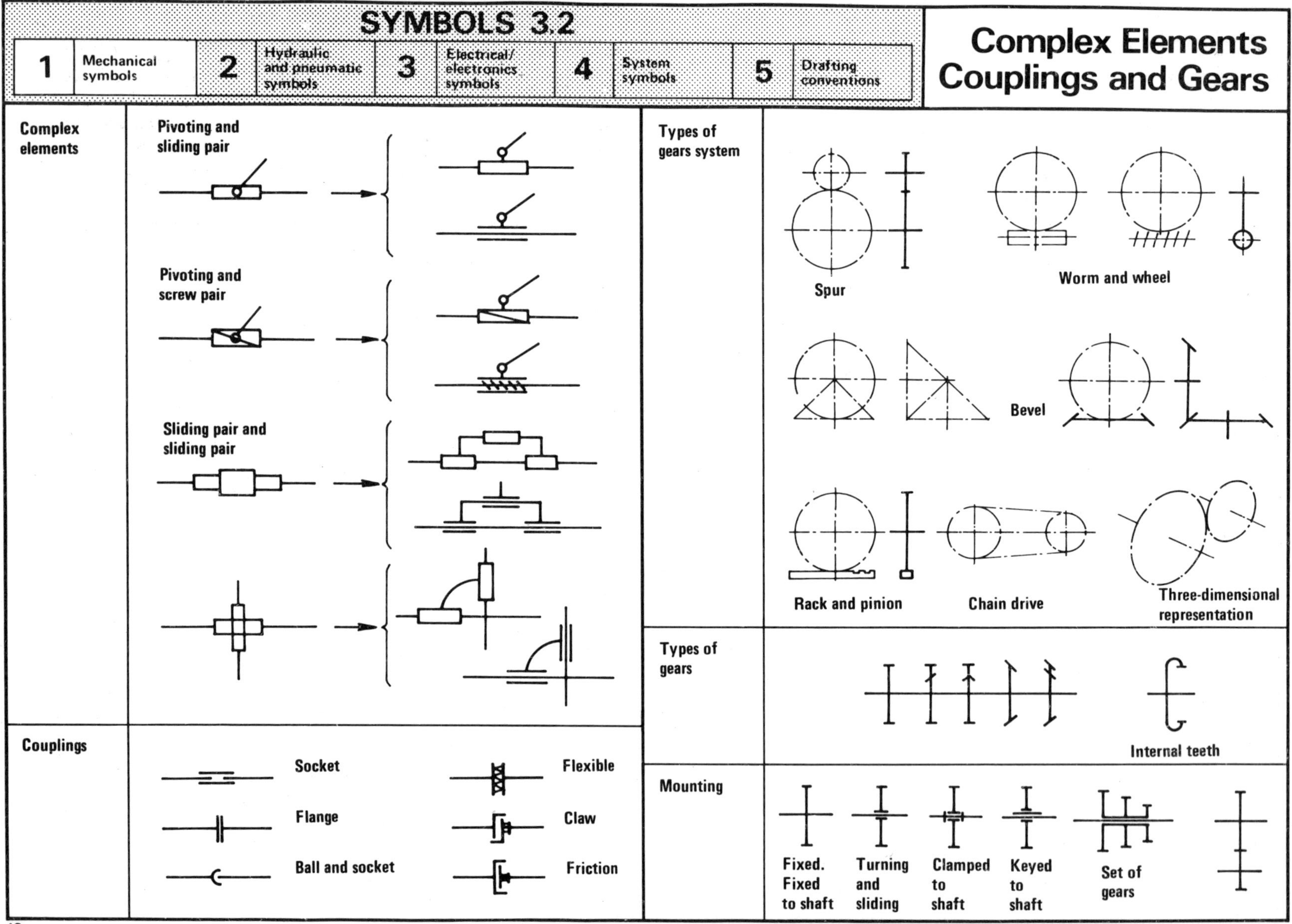
SYMBOLS 3.2
1
Mechanical symbols
2
Hydraulic and pneumatic symbols
3
Electrical/ electronics symbols
4
System symbols
5
Drafting conventions
Complex Elements
Couplings and Gears
Complex elements
Pivoting and sliding pair
Pivoting and screw pair
Sliding pair and sliding pair
Couplings
Socket
Flange
Ball and socket
Flexible
Claw
Friction
Types of gears system
Spur
Worm and wheel
Bevel
Rack and pinion
Chain drive
Three-dimensional representation
Types of gears
Internal teeth
Mounting
Fixed. Fixed to shaft
Turning and sliding
Clamped to shaft
Keyed to shaft
Set of gears

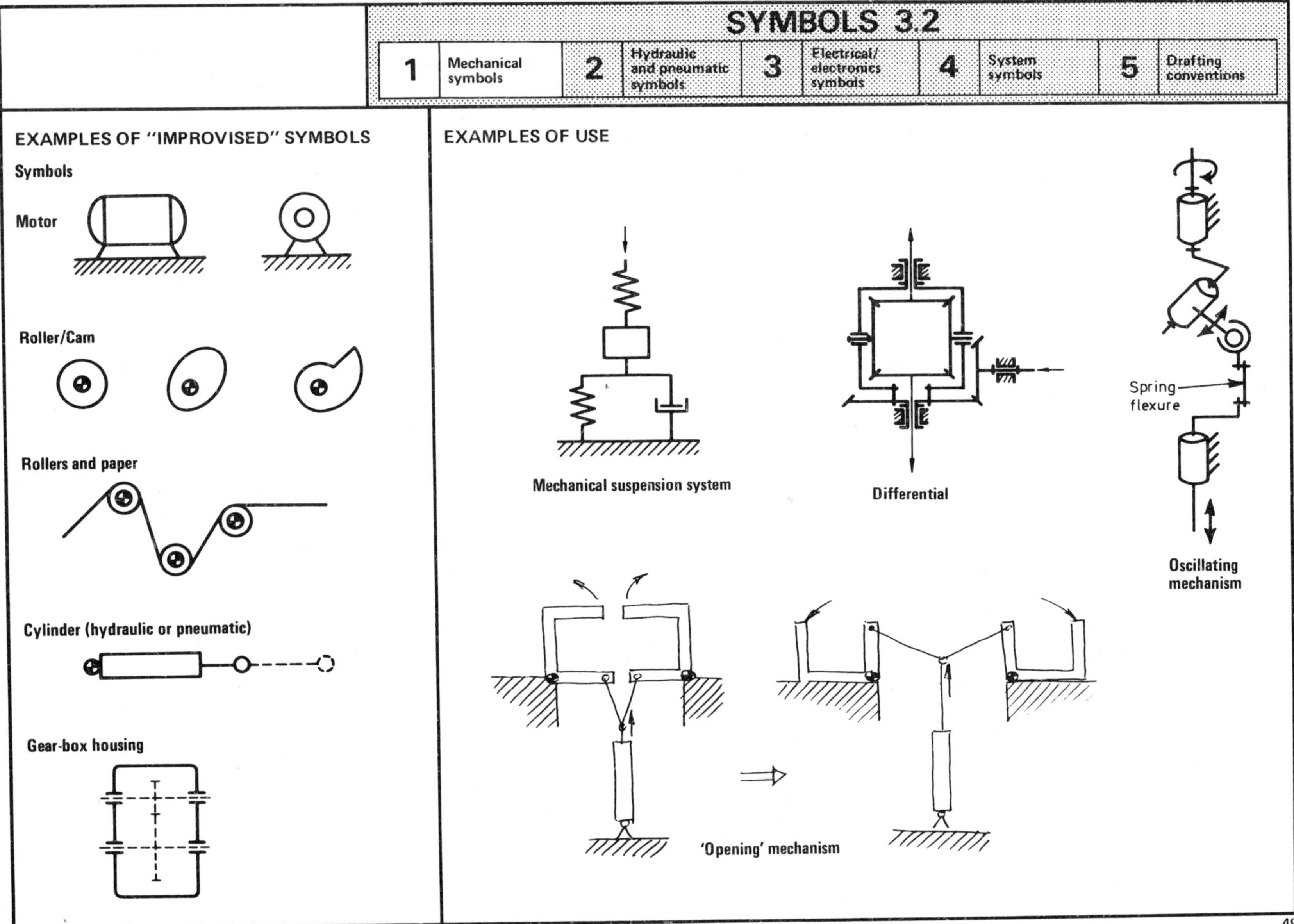

Mechanical suspension system

Differential

Oscillating mechanism

'Opening' mechanism

Hydraulic and Pneumatic Symbols

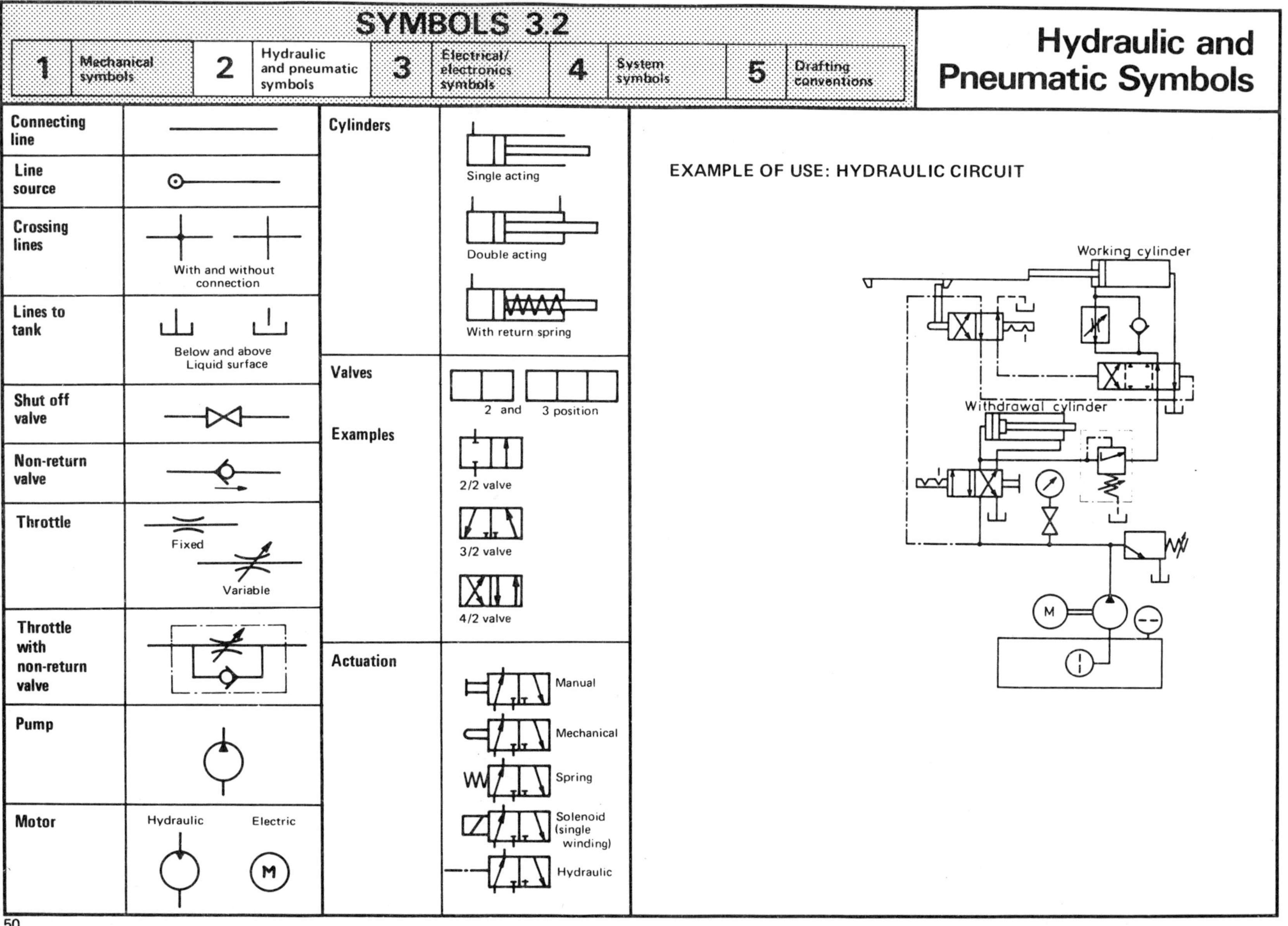

Electrical/Electronics Symbols

EXAMPLES FROM BS 3939

Symbol	Example	Symbol	Example
Conductor		Capacitor	
Crossing conductors	with and without connection	Lamp	
Frame connection		Relay coil	
Earth		Relay	CHANGE OVER
Supply	direct current; battery accumulator; alternating current		
Resistor	R; 500 Ω; non reactive; variable	Diode	pn diode
Impedance	Z	Transistor	pnp; npn
Inductor	with core	Fuse	
Transformer	with core	Switch contact	
		Meter	A; mV

EXAMPLE OF USE: CIRCUIT DIAGRAM

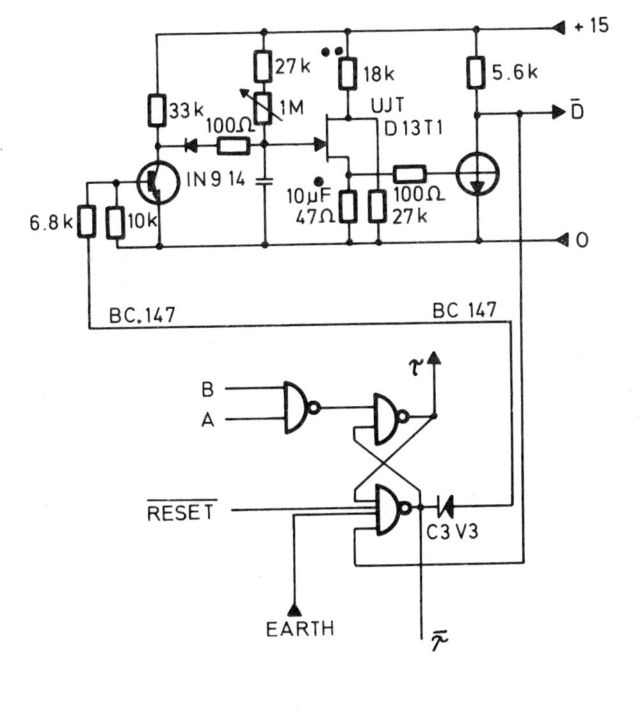

System Symbols

EXAMPLES FROM BS 4058 (1S0/R1028)

Symbol name		Symbol name	
Connecting line (data flow)		On-line storage	
Process or operation		Document	
Decision		Punched card	
Preparation or modification		Deck of cards	
Predefined process i.e. decided elsewhere e.g. sub programme		Paper tape	
Manual operation		Magnetic tape	
Auxiliary operation		Connector	
Manual input		Start, stop, halt, delay etc.	
Input/output			

EXAMPLES OF USE: A section from a flow chart in electronic data processing

Term

Engineering drawing is a standardised drawing type, recognised by the use of: first and third angle projection, set rules for the line thickness, sections and the symbolic representation of frequently recurring details.

Use

Engineering drawings are used to specify products and mechanical systems, so that they may be produced by skilled craftsmen.

A product is normally documented as a DRAWING SYSTEM, comprising:

PARTS LIST. Lists of all the components in a system and the appropriate part numbers.

ASSEMBLY OR GENERAL ARRANGEMENT DRAWING. Drawings with views and sections, enabling the arrangement of the assembled parts to be shown. Each component is identified numerically and referenced in the parts list.

DETAIL DRAWINGS. These specify: form, material, dimensions and the surface qualities.

Characteristics

The rules for drawings are contained in standards e.g. British Standard 308. Similar standards apply abroad, e.g. the German standards Deutsche Industrie-Normen (DIN).

The more important rules and some practical guidance are given in the following pages.

Orthographic projection is used

This diagram shows the principle of first angle projection, third angle projection may be illustrated in a similar way.

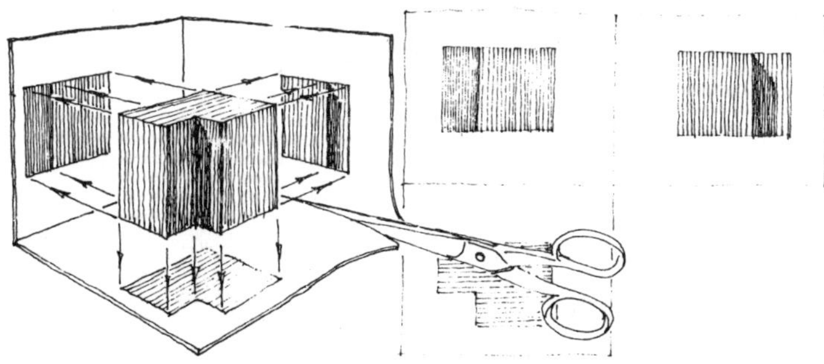

Drawing rules

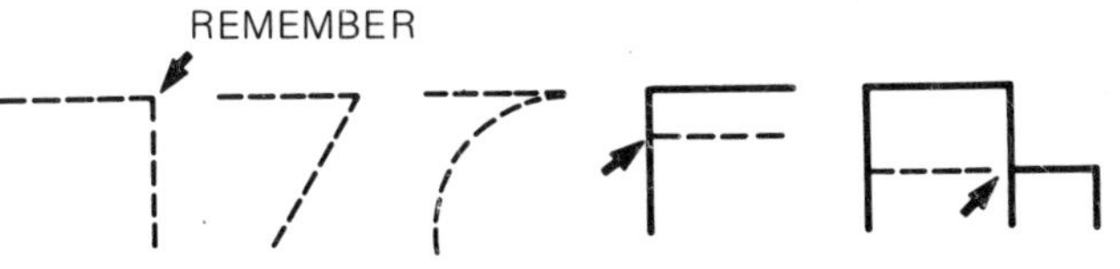

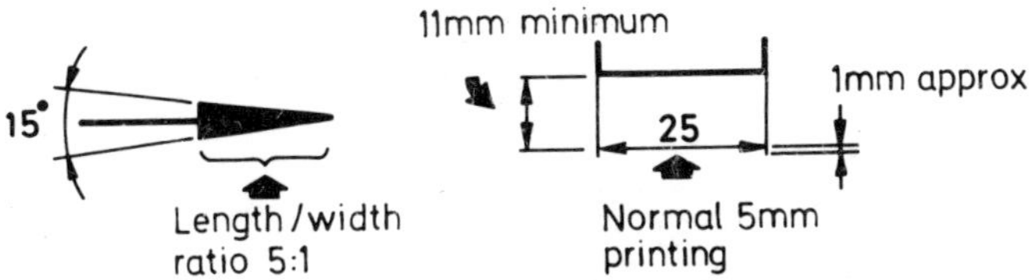

See also the following sections:
Parts list. Page 104
Assembly drawing. Pages 102–104
Detail drawing. Pages 100–101

TYPES OF LINE

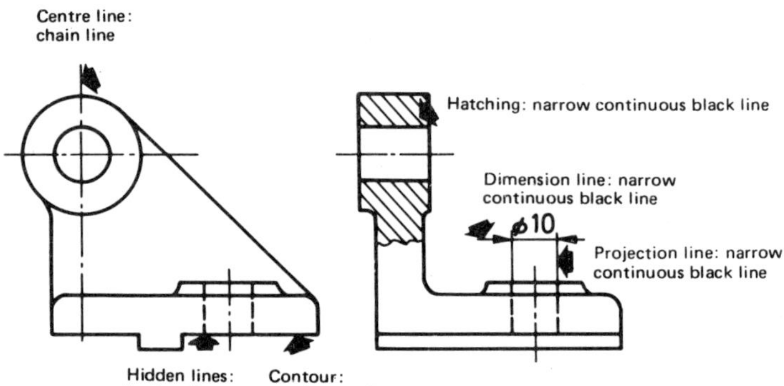

Dimensioning when space is limited

Note the positioning of the figures

Shortest dimension nearest object

Utilisation of centre line

		Relative recommended	
		Thickness	Width of line
Chain line Break line Hatch line Dimension line		1	0.3 mm
Broken line		2–2.5	0.5 mm
Bold black line (item profile)		3–4	0.7 mm

USE OF SYMBOLS

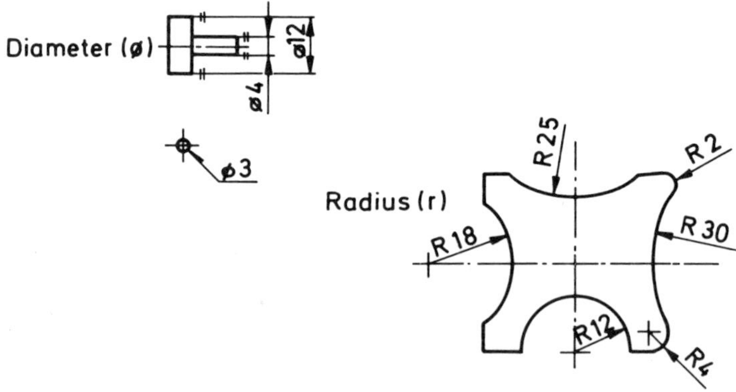

When dimensioning, consideration should be given as to how the item will be made. The selected datum should, in each case, be a convenient reference from which the craftsman can work without the need to make further calculations.

Sectioning

MARKING OF BREAKS INDICATING TYPE OF SECTION (REVOLVED SECTIONS)

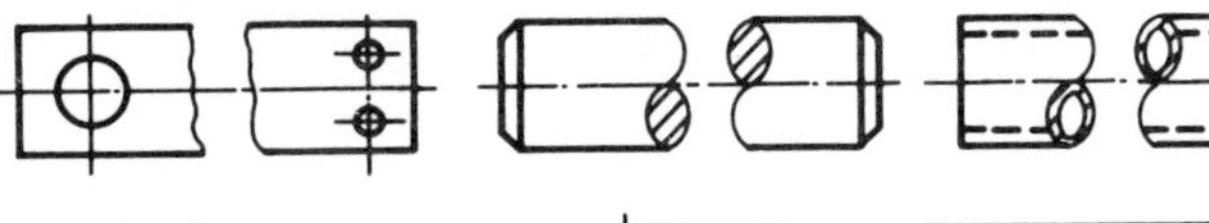

Remember: section lines are narrow, continuous and black

Remember: the main component profile lines should dominate

Partial sections are shown with cross hatch lines

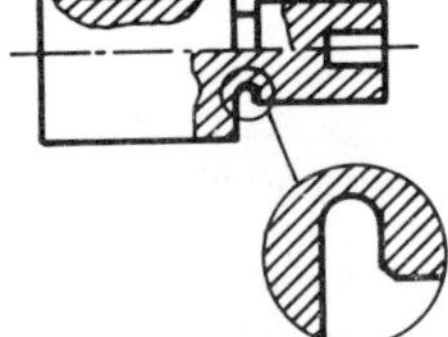

Details of sections may be shown to a different scale

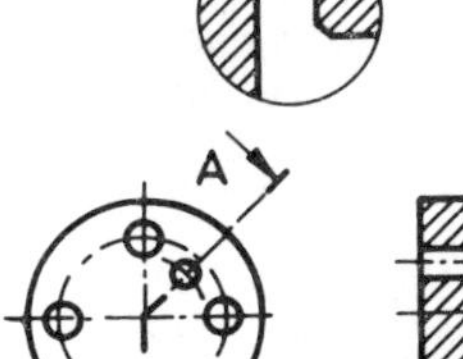

Sections can be rotated into a single plane

SPECIAL VIEWS

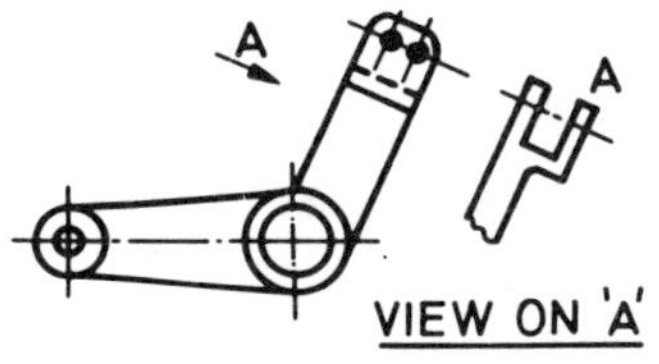

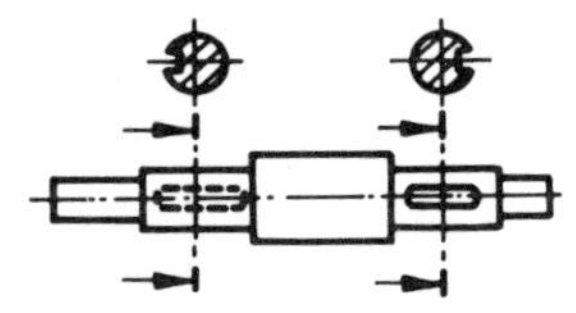

ITEMS WHICH ARE NOT SECTIONED

Solid cylindrical parts

Standard items

Balls in bearings

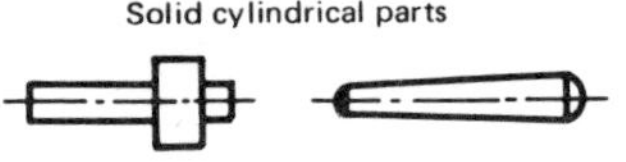

Bolts, flanges and webs

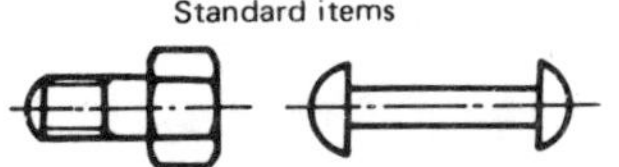

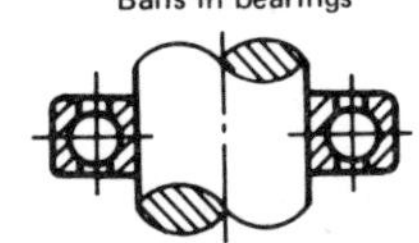

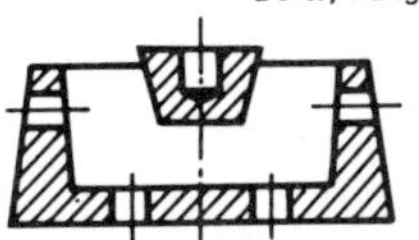

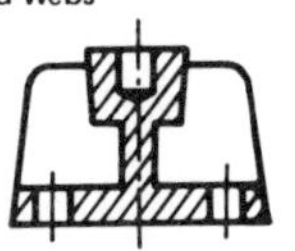

SECTIONING RULES

Same piece, same hatching
Common interfaces are drawn with one line

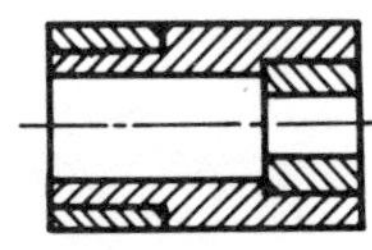

Remember: hatching is a narrow black line at 45°

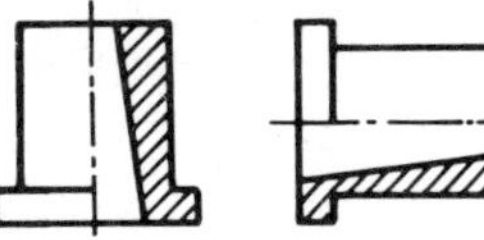

Remember: a normal centre line divides the sectioning and exterior view

Sections are defined by
- a chain line
- arrows
- section identifiers

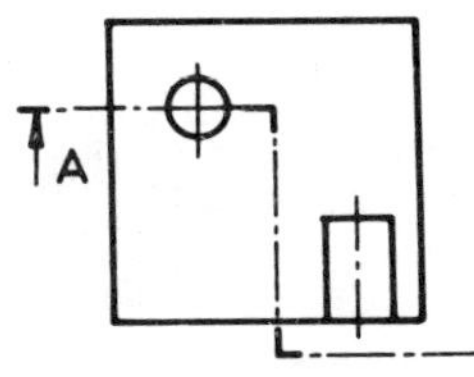

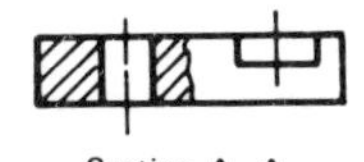

Section A–A

DIMENSIONING SECTIONS

Half dimensions for diameters

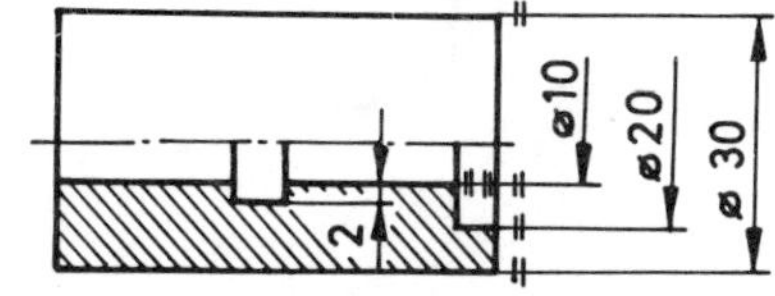

3.3 THE DRAWING CODE

TYPES OF PROJECTION

In order to represent a three dimensional solid object on a two dimensional medium, such as a sheet of paper, it is necessary to adopt a projection convention, i.e. a given point on the object is represented by a given point on the paper, this transfer being according to set rules. The following projections are used:–

3.1 Orthographic
3.2 Isometric
3.3 Dimetric
3.4 Trimetric
3.5 Oblique
3.6 Single-point perspective
3.7 Two-point perspective
3.8 Three-point perspective

(3.2–3.5: 'Pictorial' projections)

PERSPECTIVE

In addition non-conventional representations may be used, particularly when freehand sketching, which become a subconscious mixing of the different types of projection. This is possible because of the similarity between the construction rules, in some cases these are the same. See also the procedure sheet on page 74.

FIRST ANGLE

THIRD ANGLE

Isometric

Dimetric

Trimetric

Oblique

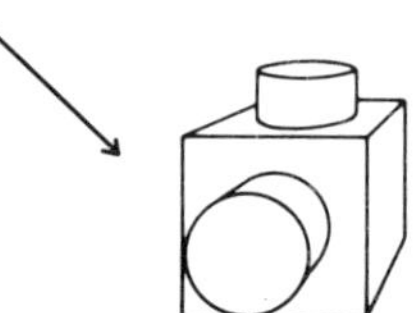

Single point perspective

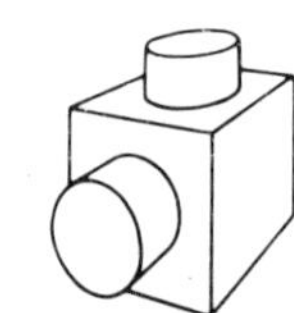

Two point perspective

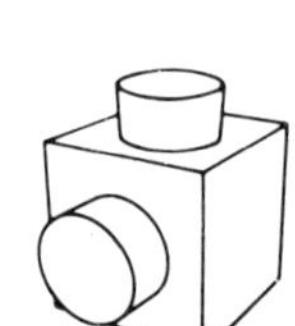

Three point perspective

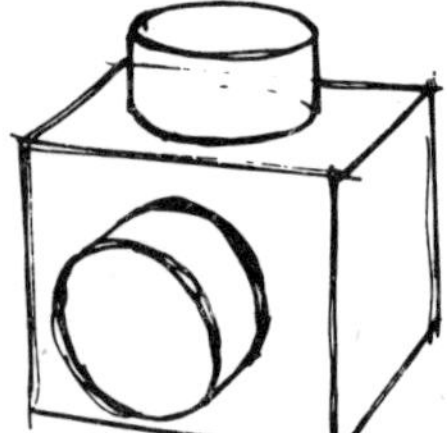

TYPE OF PROJECTION

?

TYPES OF PROJECTION 3.3

1	Orthographic projection	2	Pictorial projection	3	Oblique projection	4	Perspective

Orthographic Projection

Term

Orthographic projection

Use

Orthographic projection is widely used in the technical specification of items or products for manufacturers and other purposes, it also conforms to certain standard conventions.

Characteristics

The basis of orthographic projection is shown below. The view is projected onto the picture plane and, in this example, the picture plane is rotated through 90°, about the line XX, to form a single plane with the vertical picture plane. (N.B. In the example for third angle projection the picture plane is, in effect, viewed from behind.) Normally views are taken in three mutually perpendicular directions to represent the object fully.

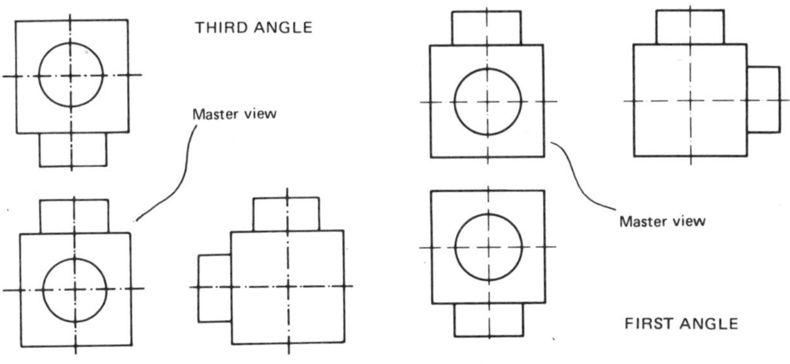

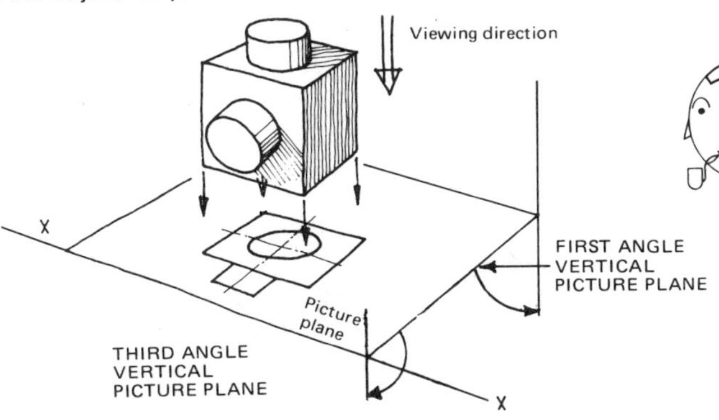

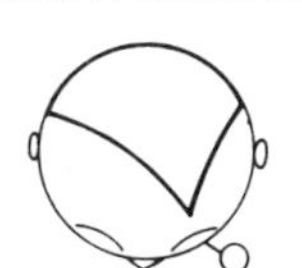

A simple way to differentiate between first and third angle projection is to remember the following rules: (refer to the drawings of the man's head):

(i) first angle – the view is drawn on the opposite side of the master view from the view point.

(ii) third angle – the view is drawn on the same side of the master view as the view point.

Isometric Projection

Term

Isometric projection, isometric drawing.

Use

Isometric projection is often used for three dimensional drawings, e.g. technical illustrations, because the angles of the axes and the proportions are simple to use.

Characteristics

The main axes (in a Cartesian co-ordinate system) are represented by lines forming mutual angles of 120°.

The proportions along the axes are in the ratio 1:1:1. The name isometry indicates that the proportion along each of the axes is the same. An isometric scale may be used if it is important that the drawn object appears the correct size.

Circles having centre lines parallel with the main axes are drawn as ellipses with the major axis at right angles to the isometric drawing of the centre line. The proportion of the axes is 1:0.58.

DRAWING OF A CUBE

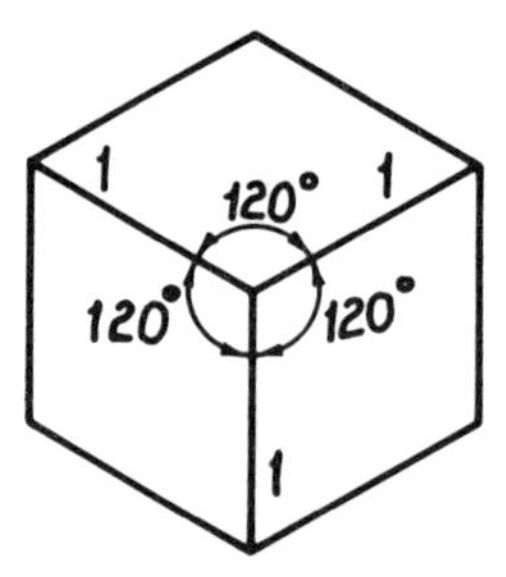

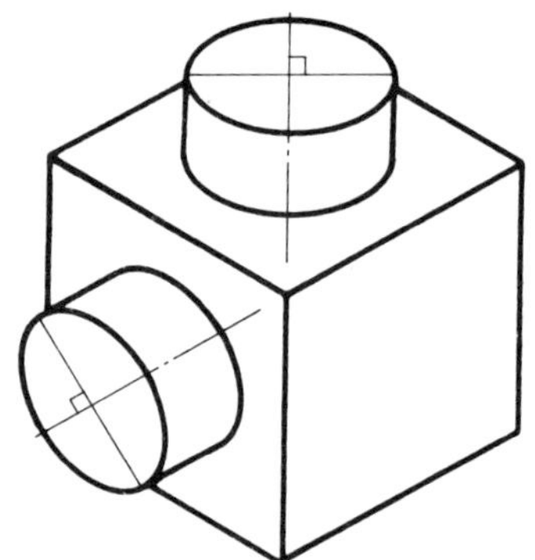

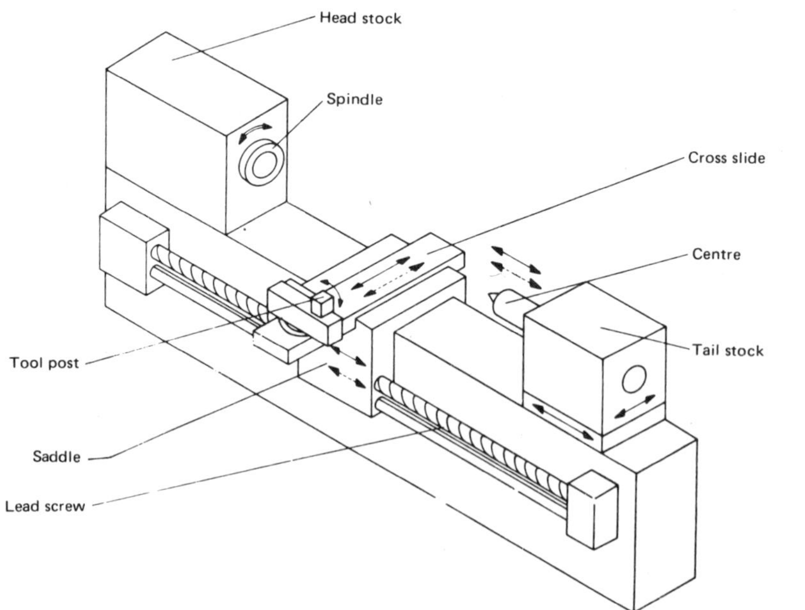

Dimetric Projection

Term

Dimetric projection, dimetric drawing.

Use

Dimetric projection is often used for three dimensional representation. If convenient proportions are selected construction is simple and the pictorial effect good.

Characteristics

The main axes (in a Cartesian co-ordinate system) are often drawn to form angles of: 42°, 90° and 7°, with the horizontal.

Proportions along the axes for the above angles would be 1:1:½. The term dimetric indicates that the proportions of two of the axes are the same. The proportions of the third axis depend on the angles of the axes.

Circles on centre lines parallel to the main axes are represented by ellipses, the major axis makes a right angle with the projected centre line.

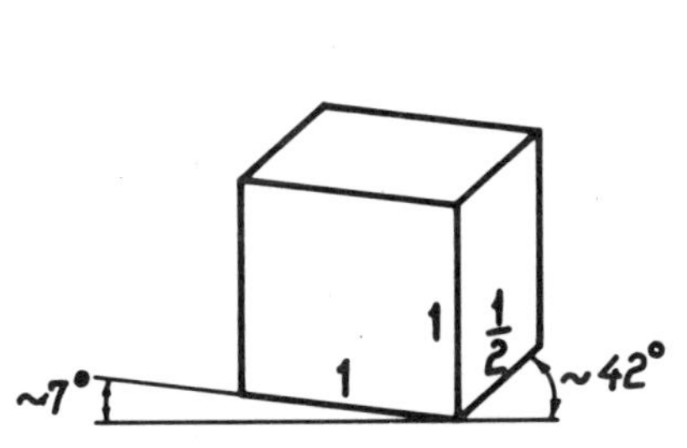

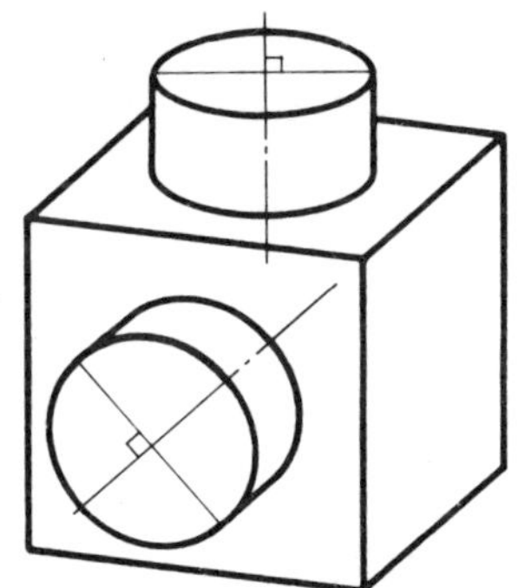

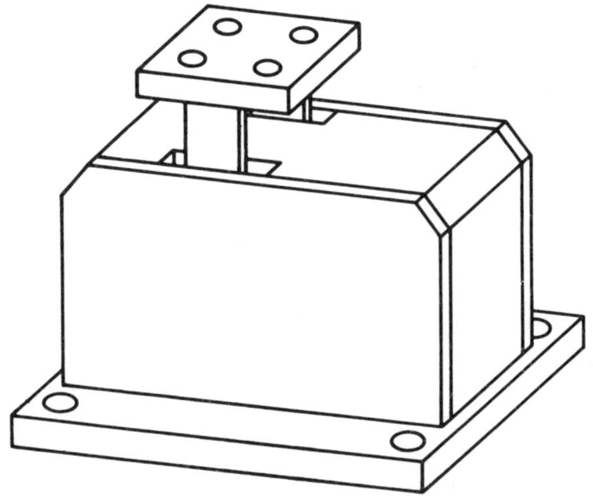

Trimetric Projection

Term

Trimetric projection, trimetric drawing.

Use

Trimetric projection can give good proportions in a three dimensional representation, but as there are three different scales on the axes, it becomes more complex than isometric and dimetric.

Characteristics

With trimetric projection, the angular positions of the main axes can be chosen freely, however the proportions between the axes depends on this choice. In trimetric projection, the circles are drawn as ellipses with their major axes at right angles to the projected centre lines.

Isometric and dimetric projections are special cases of trimetric projection.

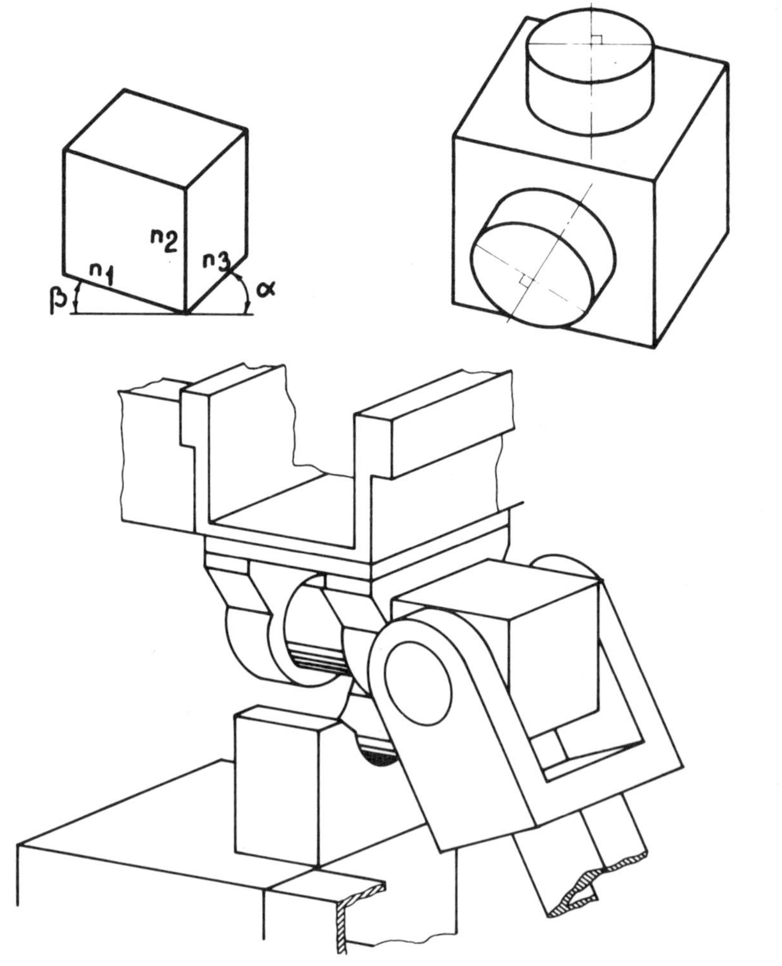

Representing Circles

The projections: ISOMETRIC, DIMETRIC, TRIMETRIC etc, require circles on their 'inclined' faces to be represented as ellipses. Each ellipse must have its major axis at right angles to the normal of the plane in which the ellipse lies. This may be illustrated by the following reminder:

The technique of freehand sketching ellipses should be developed and experience gained in drawing three dimensional representations. The starting point is a sketch of a cube, with little regard being paid to the formalised projection requirements.

THE RULE THAT THE MAJOR AXIS OF THE ELLIPSE MUST BE AT RIGHT ANGLES TO THE NORMAL OF THE PLANE CONTAINING THE ELLIPSE ALWAYS APPLIES FOR ISOMETRIC/DIMETRIC/TRIMETRIC PROJECTIONS AND SHOULD ALSO BE USED WHEN AXES/ANGLES ARE ARBITRARILY CHOSEN.

This rule is often sinned against:—

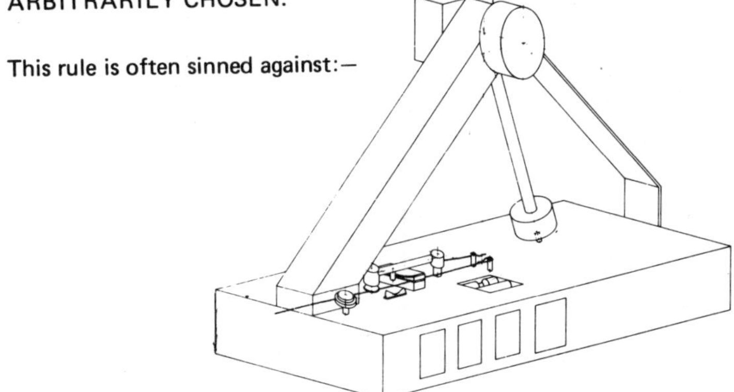

ISOMETRIC AND RELATED PROJECTIONS 3.1-3.4

A true circle is most conveniently projected in the following way. Enclose the circle in a square such that the sides of the square just touch the circumference of the circle. The square box is then drawn in the appropriate pictorial projection, and provides the enclosure for the ellipse. This technique is sometimes called boxing.

To sketch circles freehand without an imaginary or physical reference to a 'box' can produce peculiar results:

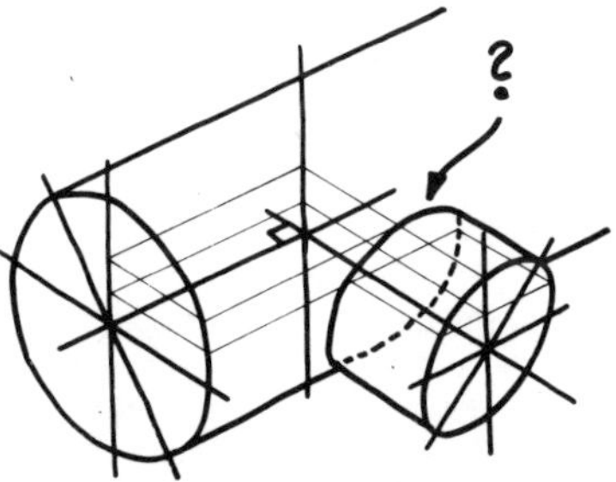

As a check the ellipses in this example can be boxed as shown below. It is then seen that the two ellipses are not tangential to the projected 'squares' in which the original circles were boxed.

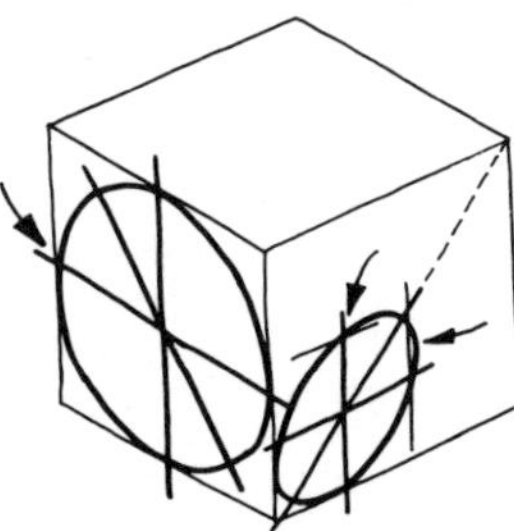

IT IS RECOMMENDED THAT AS A CHECK ON THE CORRECTNESS OF THE PRESENTATION, BOTH THE BOX AND ELLIPSE ARE DRAWN ON THE ISOMETRIC PROJECTION.

Oblique Projection

Term

Oblique projection.

Use

Oblique projection is used for pictorial presentation, e.g. technical illustrations.

Characteristics

Oblique projection has the advantage that figures and curves lying in planes parallel to the picture plane are drawn true to size. It is therefore important to take advantage of this characteristic by drawing complex shapes so that they lie in the picture plane.

The three main axes are drawn horizontal, vertical and at an angle U to the horizontal, U may be chosen freely, but is often 45° or 30°.

Dimensional proportions along the axes: 1:1:n, where n is normally chosen to be ½≤n≤1; often n = ½.

Circles being in planes parallel to the picture plane, are drawn as true circles. All others are drawn as ellipses, the major axes of which are *not* normal to the centre line of the view.

DRAWING OF A CUBE

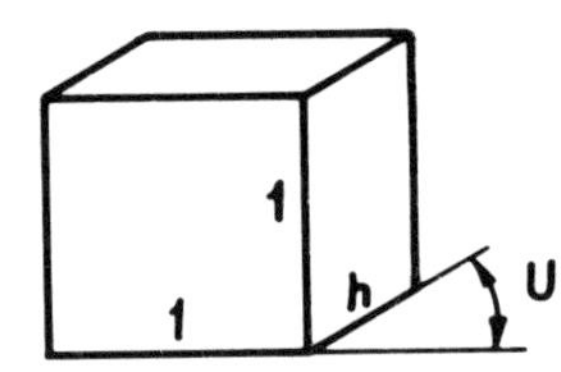

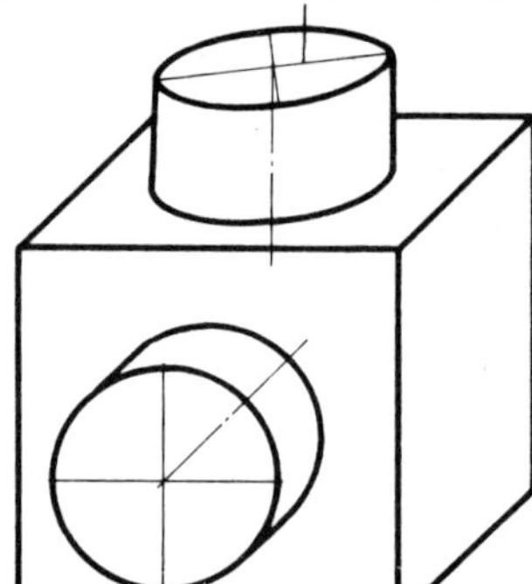

Examples

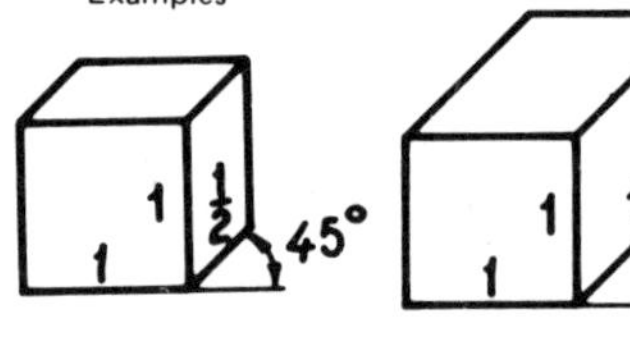

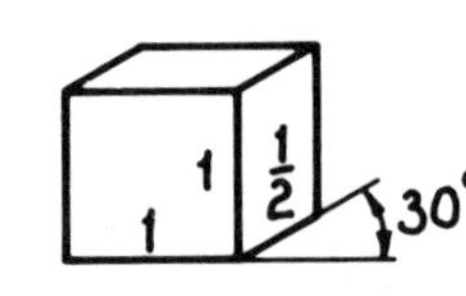

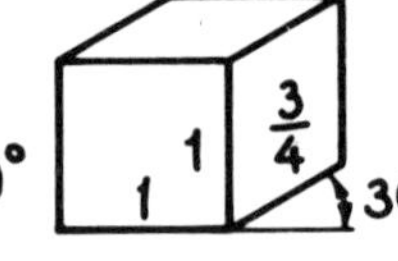

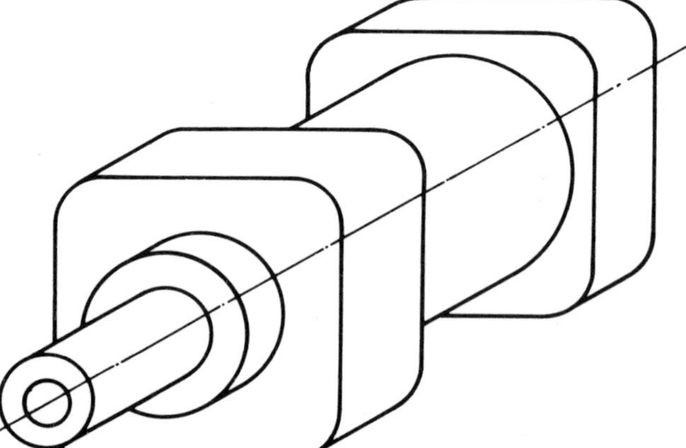

Single-Point Perspective

Term

Single point perspective, perspective with a single vanishing point.

Use

Single point perspective is used to add realism to three-dimensional representations, e.g. freehand sketching and technical illustrations. It is the easiest of the three types of perspective to apply. However, it does not always give reasonable proportions to the view.

Characteristics

The following rules apply to all forms of perspective. Parallel lines are drawn in bundles through the same point, i.e. the vanishing point. Lines in planes parallel to the picture plane are however, drawn with their true relative inclinations.

For single point perspective. The principal axes in Cartesian co-ordinates are drawn horizontally, vertically and as a line through the point H on the horizon h. H is the vanishing point for the third axis.

Proportions. Dimensions are in proportion for the vertical and horizontal directions in the frontal plane, but diminish in size as the distance from the picture plane increases.

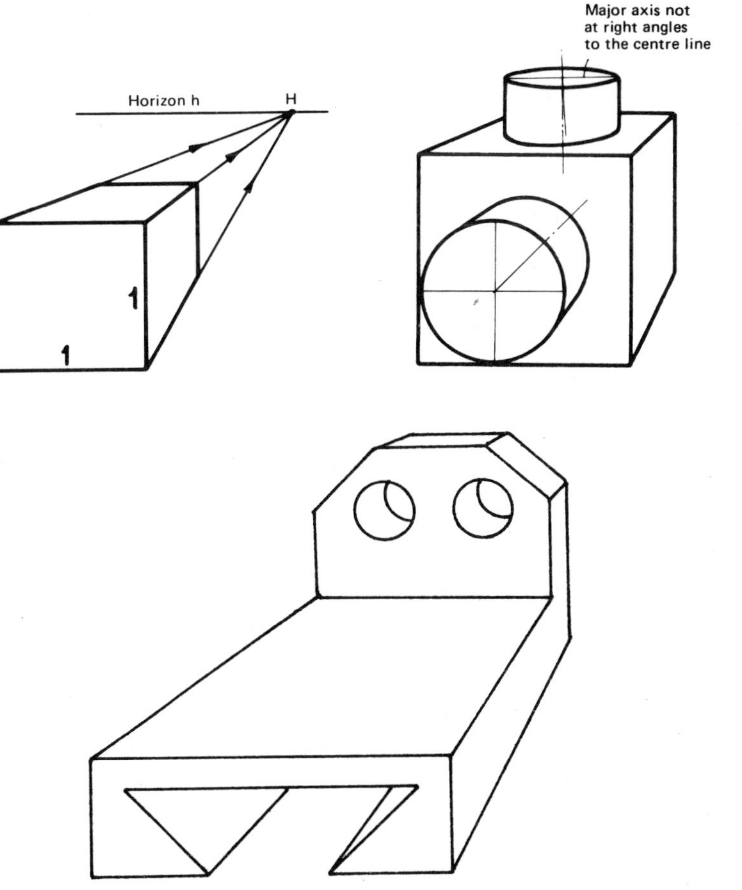

Two-Point Perspective

Term

Two-point perspective, corner perspective.

Use

Two-point perspective is used for pictorial drawings, where it is important that the view is a good likeness to the object. This type of presentation is used extensively in freehand sketching, and for technical illustrations.

Characteristics

The following rules apply to all forms of perspective. Parallel lines are drawn in bundles through the same point; i.e. the vanishing point. Lines in planes parallel to the picture plane are however, drawn with their true relative inclinations.

For two-point perspective. The principal axes in Cartesian co-ordinates are drawn; vertically, and as two lines, each of which passes through its own vanishing point on the horizon. All vertical lines are drawn as vertical lines. N.B. this is not the case for three-point perspective.

Proportions. Dimensions are in proportion along vertical lines, but diminish with increasing distance from the picture plane.

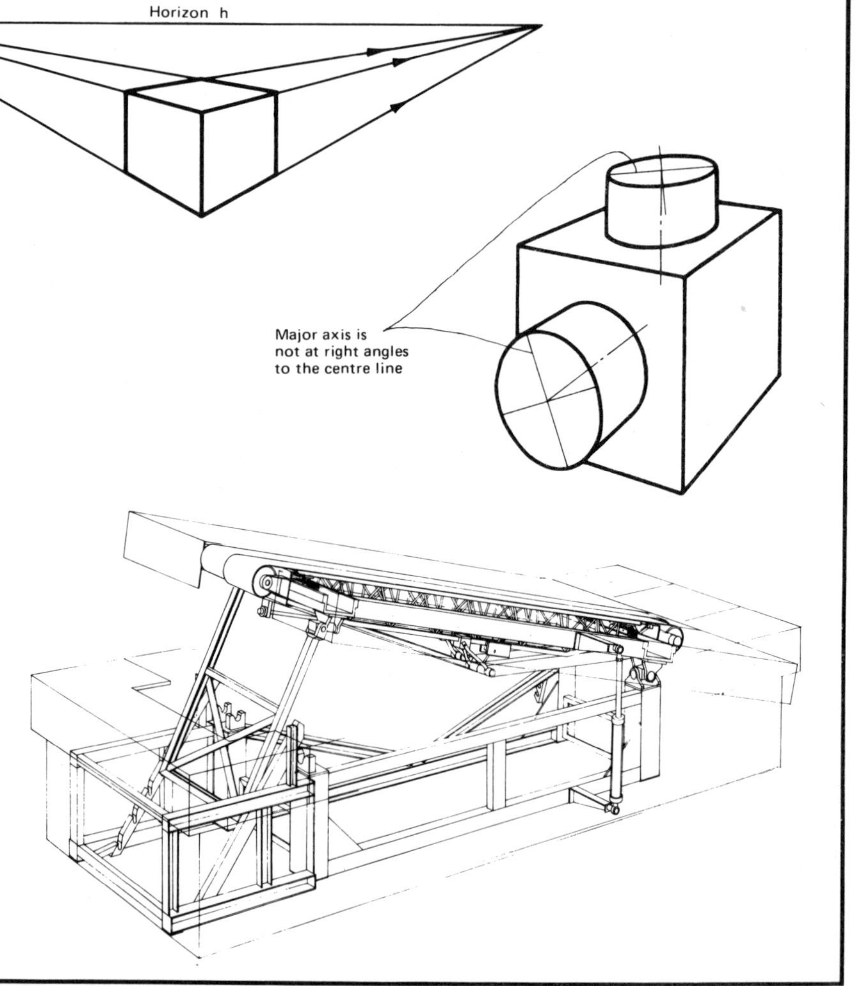

Three-Point Perspective

Horizon h

Term

Three-point perspective, bird's eye perspective (high view point), worm's eye perspective (low view point).

Use

Three-point perspective is used for three-dimensional presentation, where the emphasis is on an extremely good likeness to the object and particularly where there is a need to emphasise the size of the object. It is very suited to freehand sketching, but requires practice. Its construction is complex and it is therefore only used for technical illustrations, where the situation warrants the extra work.

Characteristics

In three-point perspective, the object is drawn as seen by the eye. The principal Cartesian co-ordinate axes are represented by three lines, each passing through its own point of direction. All vertical lines pass through a point of direction and are therefore vertically inclined.

Proportions: None of the axes are drawn in proportion.

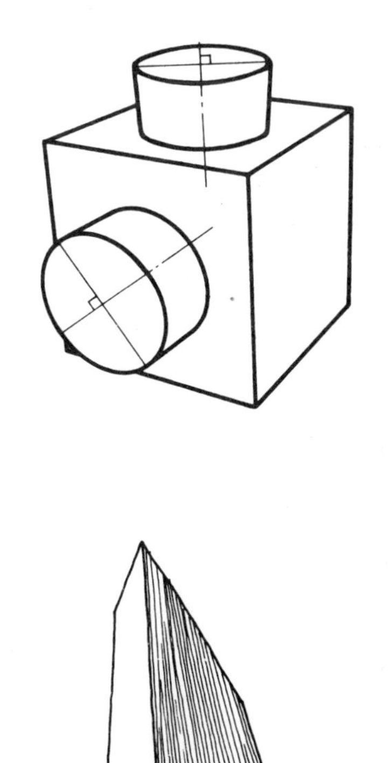

4. DRAWING TECHNIQUE

4.0 DRAWING TECHNIQUE

DRAWING TECHNIQUE is the method used to create the elements of a drawing: the lines.

In design they cover the spectrum from freehand sketching to automatic plotter drawing, and include the use of straight edges and templates.

Individual techniques are not clearly divisible between the following six sections. N.B. some of the freehand sketching techniques can be combined with other techniques.

CHOICE OF TECHNIQUE is made with the drawing's use in mind. Certain choices are obvious, e.g. layouts cannot be drawn to scale freehand. In other situations, the cost of producing 'exotic' drawings must be assessed against any benefits in improved communication. Acceptable workshop drawings can often be produced freehand, but tradition may mean that workshop drawings have to be produced on a draughting machine with a careful drawing technique.

The following aspects will now be discussed: uses, drawing aids, stages of drawing (progress in time), techniques for drawing lines, drawing methods e.g. by hand, constructions, and drawing rules.

FREEHAND SKETCHING 4.1

Use

Freehand sketching is used in design for modelling, communicating, and documenting and may employ the use of block diagrams, sketches of principle, structure and form, and preliminary working drawings etc.

Sketching is a quick and handy tool for recording suggestions and ideas. For self communication during the choosing and solutions sketching stimulates the imagination. Drawing uncovers obscurities, creating questions, and leading to new ideas.

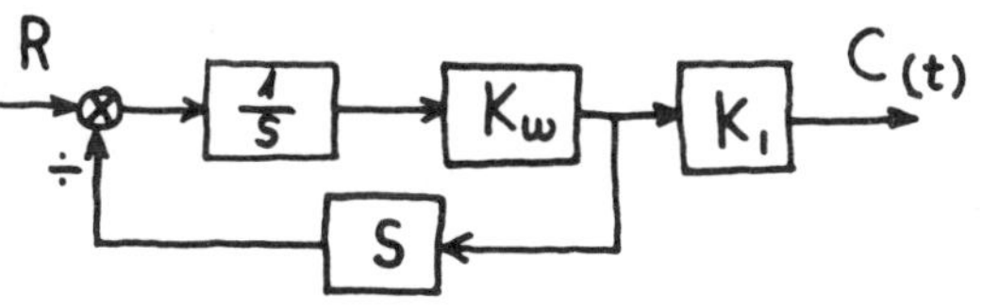

BLOCK DIAGRAM

Aids

None, apart from the drawing tool and drawing medium.
Drawing tool: Pencil, ball pen, drawing pen, felt tip pen, chalk, etc.
Medium: sketching paper, tracing paper, blackboard, flip chart, etc.

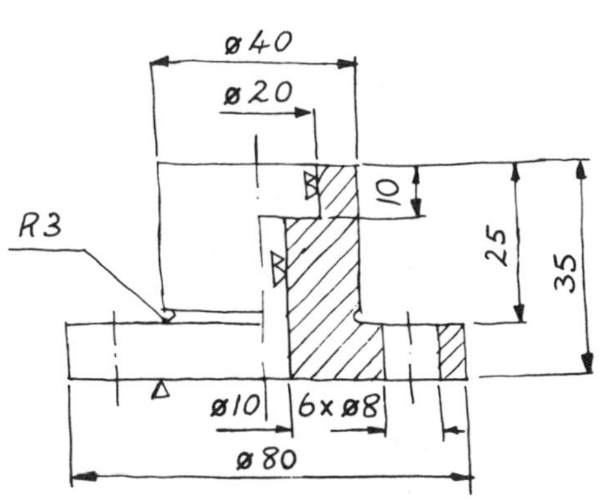

DETAIL DRAWING

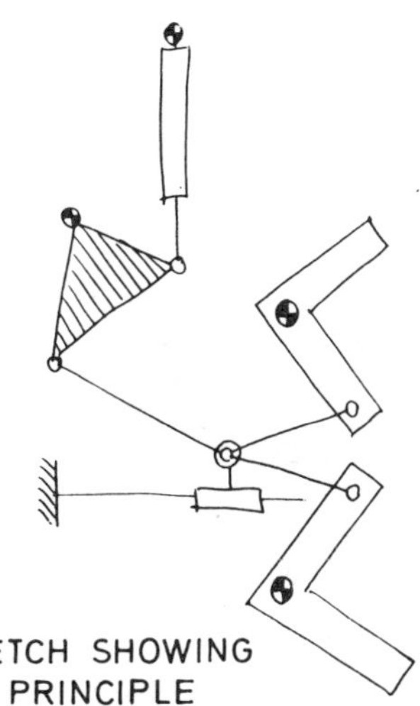

SKETCH SHOWING PRINCIPLE

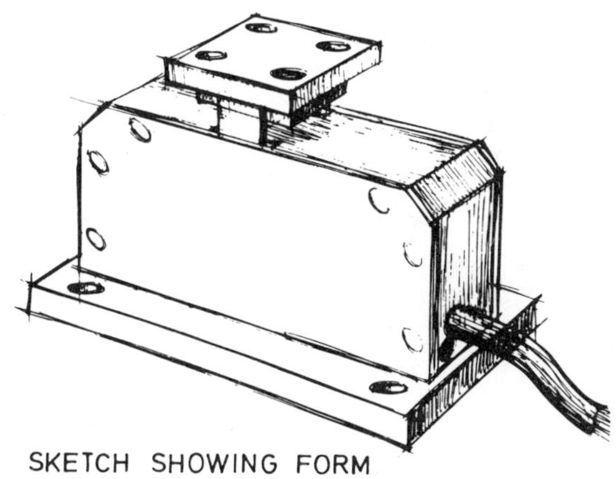

SKETCH SHOWING FORM

Stages of Drawing

It is advantageous to divide freehand sketching into four stages:

1. The principal lines are sketched to decide the position and proportions of the drawing.
2. The details are completed, aided by suitable construction lines.
3. The contours of the object are drawn in bold and black.
4. Construction lines are erased or the sketch transferred to another sheet. This may not be necessary if there is sufficient contrast between the construction lines and profile lines.

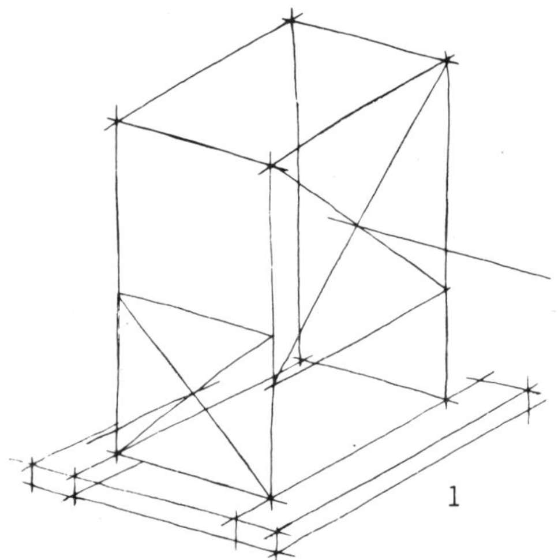
1

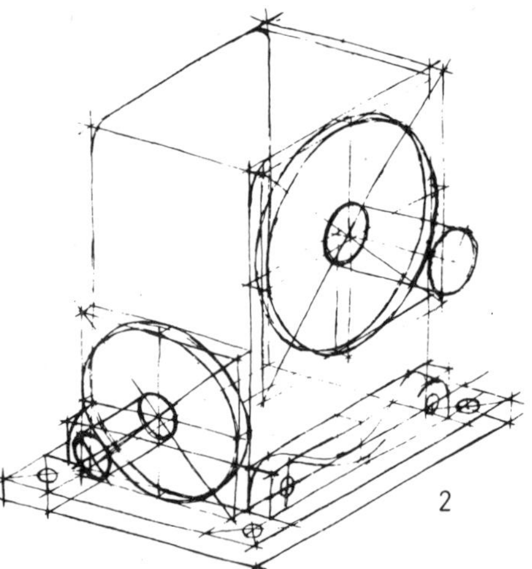
2

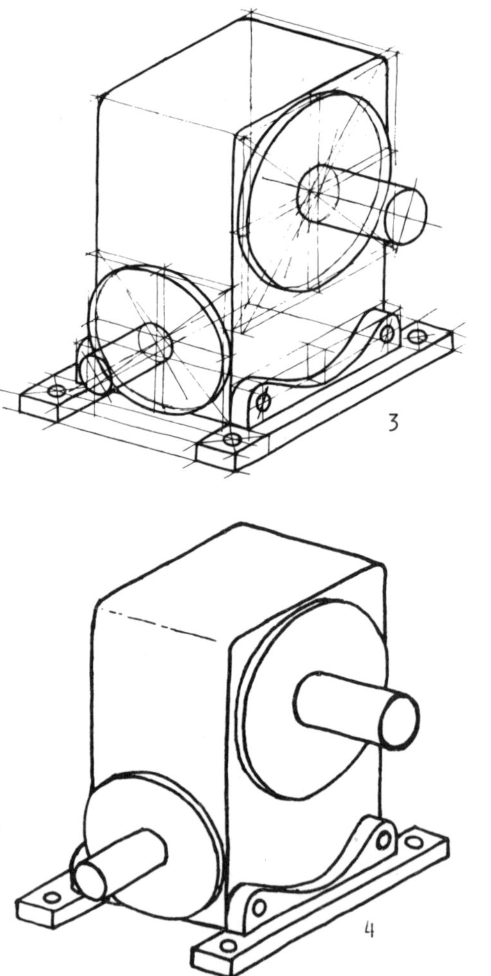
3

4

The standard of the sketch is dictated by: the number of sketches, the time available, the user, etc.

Figure 3 could be used for self-communication and Figure 4 for external use.

Hand Movement

Sketching is made easier by developing certain routines for drawing straight and curved lines using only a pencil.

It is possible to draw long lines straight by supporting the wrist, the heel of the hand or the elbow on the table.

A contrast between thickness and degree of blackness makes the use of construction lines possible. It also allows alteration without the essentials of the sketch being lost.

Examples of contrasting lines in freehand sketching.

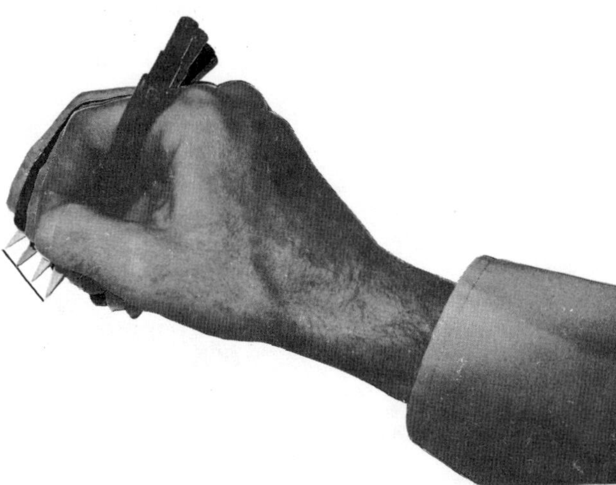

Short lines and curves are drawn by supporting the heel of the hand on the surface and only moving the fingers.

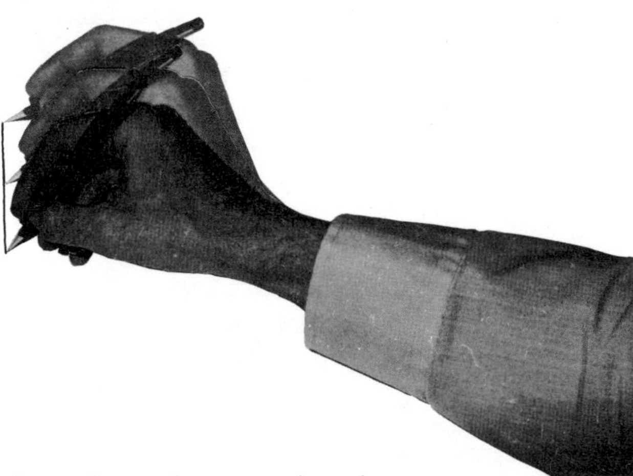

Longer lines and curves are drawn by supporting the elbow and the wrist on the surface and moving the fingers and hand (the finger movements are for correction).

Hand Movement

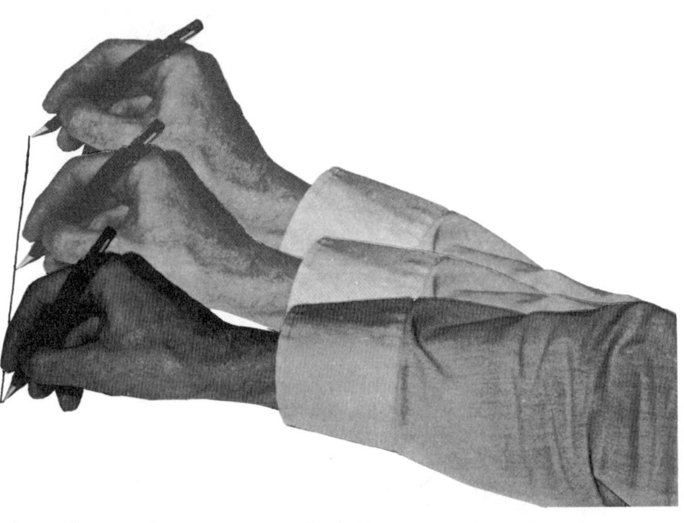

Long lines and curves are produced by supporting the elbow on the surface and moving the forearm and the fingers. (The finger movements are for correction).

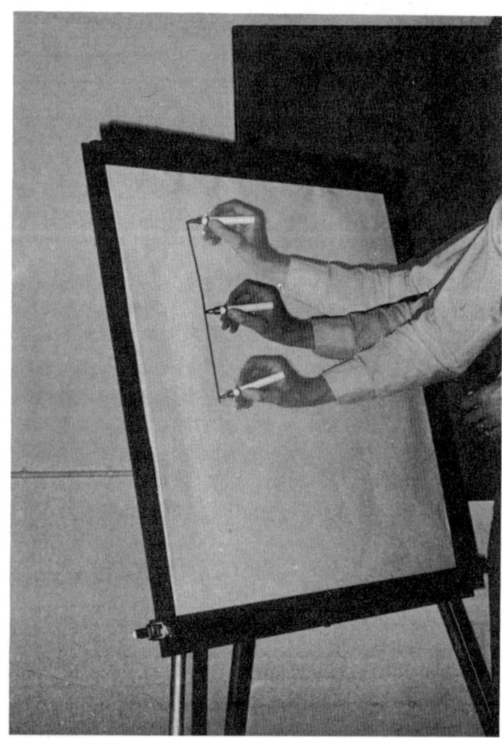

Very long lines, e.g. on a flip chart or chalk board are made by moving the arm. It is the movement of the wrist which ensures that the line is straight.

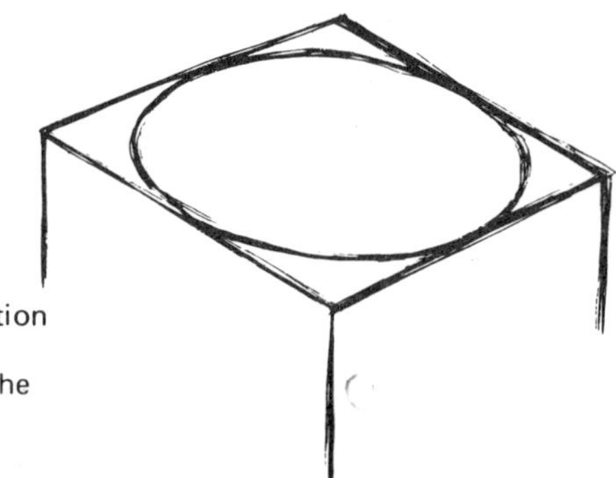

For quick sketching, time can be saved by first sketching a light line, without construction lines, and then correcting its position by gradually moving it and finally increasing the blackness by pressing harder on the pencil.

Control of the Hand

When two points are to be connected by a straight line, the position of the extreme point should be concentrated upon, and a continuous effort made to ensure that the line is always heading for this point.

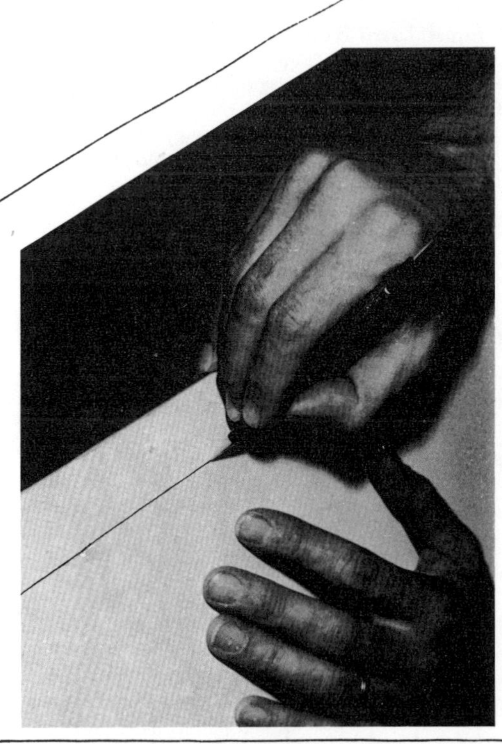

Parallel lines may be drawn by using the edge of the pad or the table as support for the hand. This also applies when the lines are some distance from the edge.

Methods of Projection

TYPES OF PROJECTION

Four types are normally used in three-dimensional drawings: isometric, dimetric and oblique projection together with perspective, see procedure sheets 3.3. These three projections cannot normally be used formally for three-dimensional freehand sketching. The advantage of the freehand sketch, i.e. quicker use of the pencil, will be lost. Spontaneous visual presentation is therefore employed, which may approximate to a formal projection, to give the best result appropriate to the user. The projection may, for example, be decided by drawing a box which appears to have the right proportions.

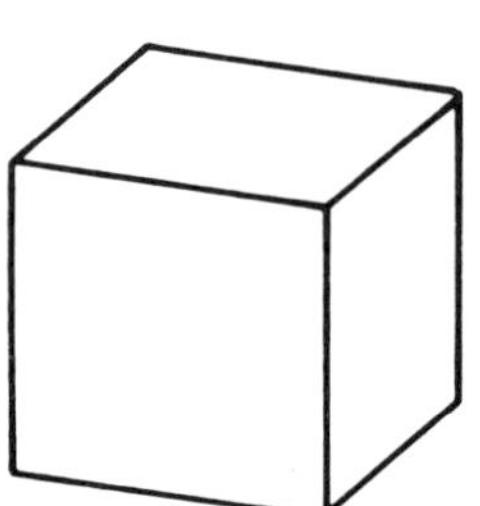

Dimetric

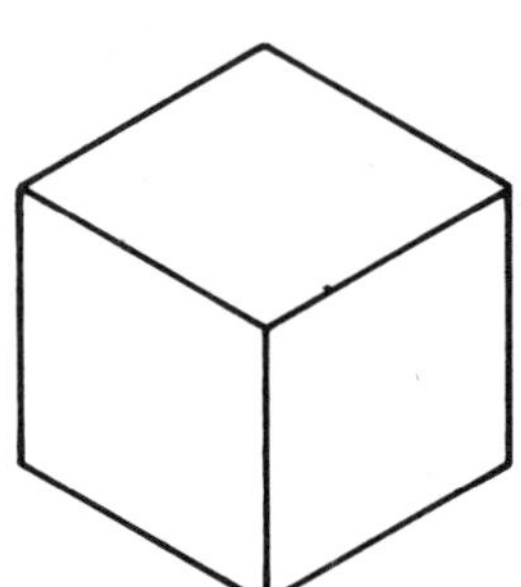

Isometric

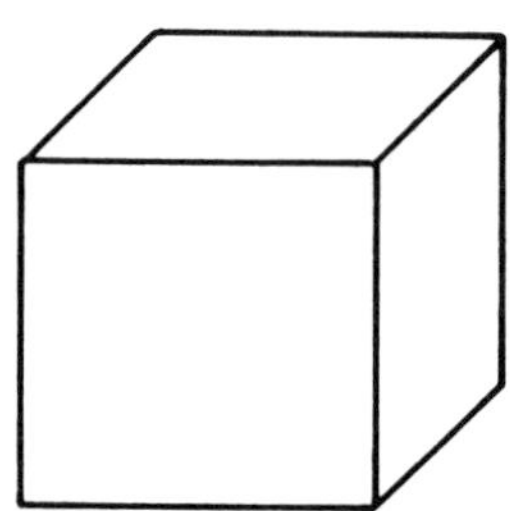

Oblique

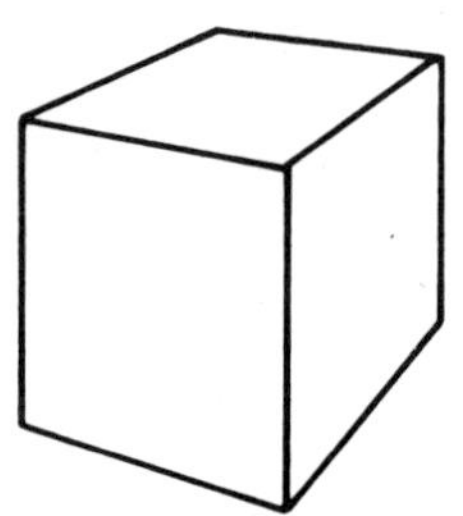

Two point perspective

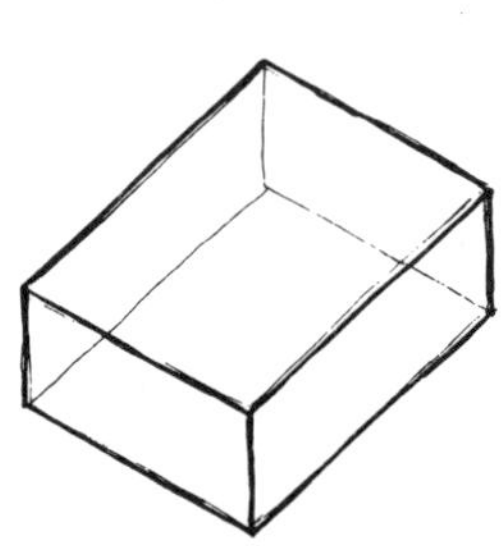

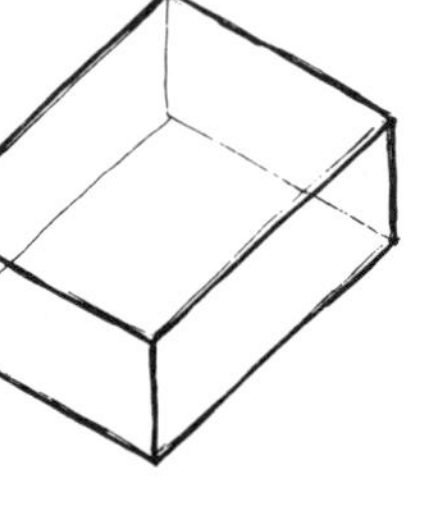

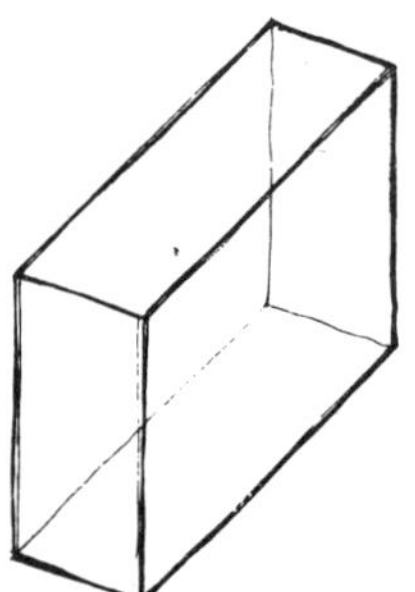

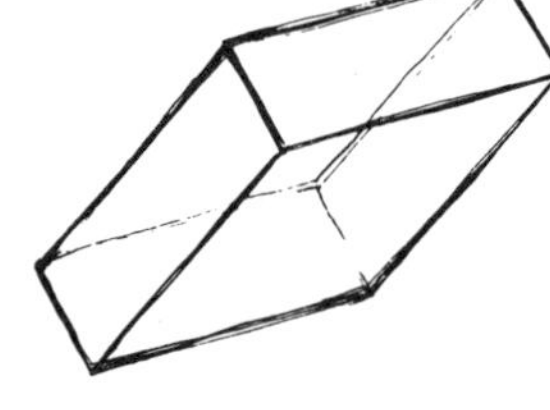

Type of projection ?

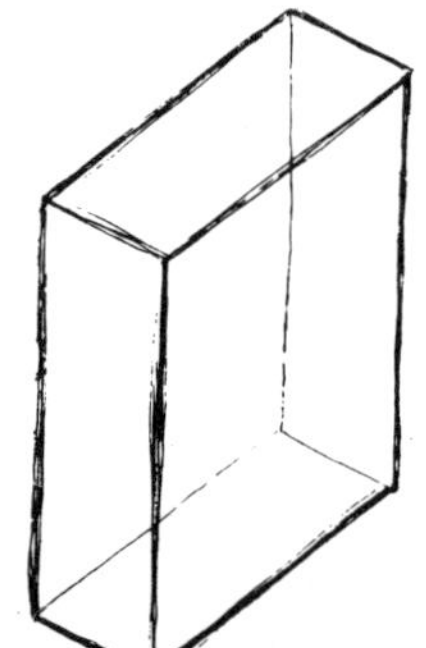

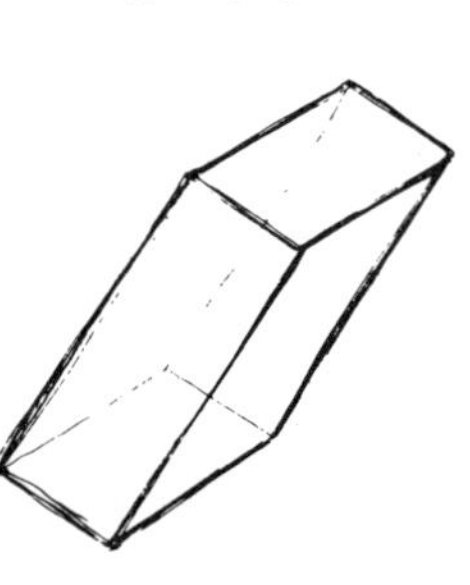

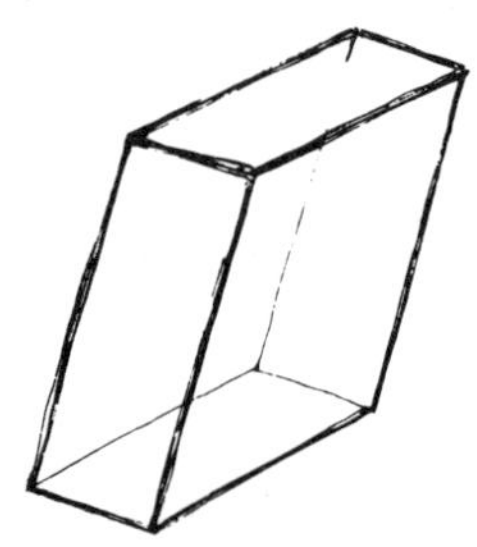

For natural presentation, perspective may be used consciously, or unconsciously.

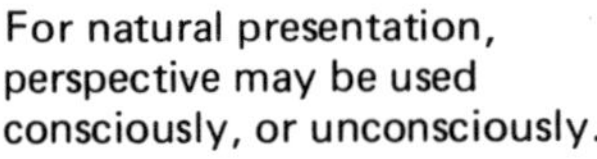

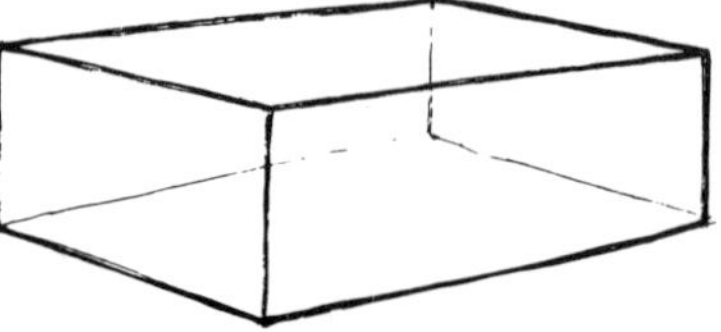

Basic Form Method

As a foundation for sketching and to ensure that proportions are correct, it is an advantage to imagine the object as being made up of approximate simple geometric figures. These basic forms which may be: boxes, cylinders, spheres, prisms etc., are sketched first. The correct sketch proportions are achieved by carefully deciding the proportions of the individual basic forms and their mutual positions.

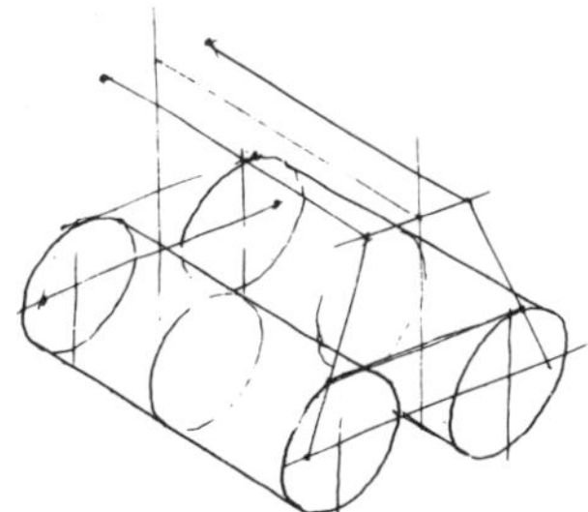

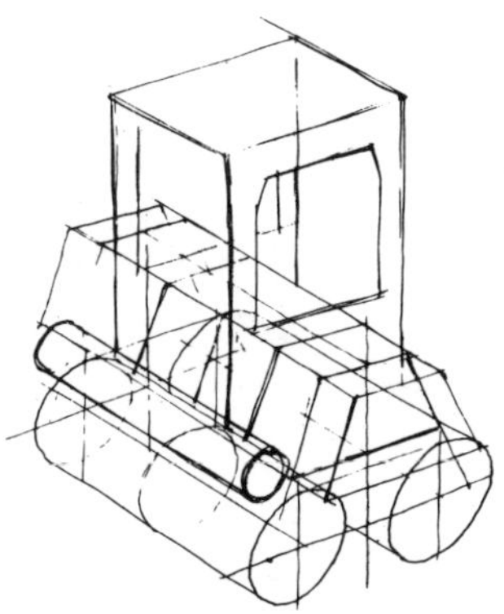

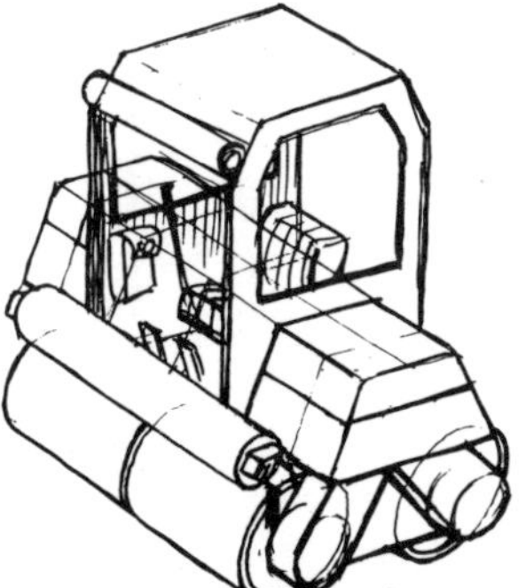

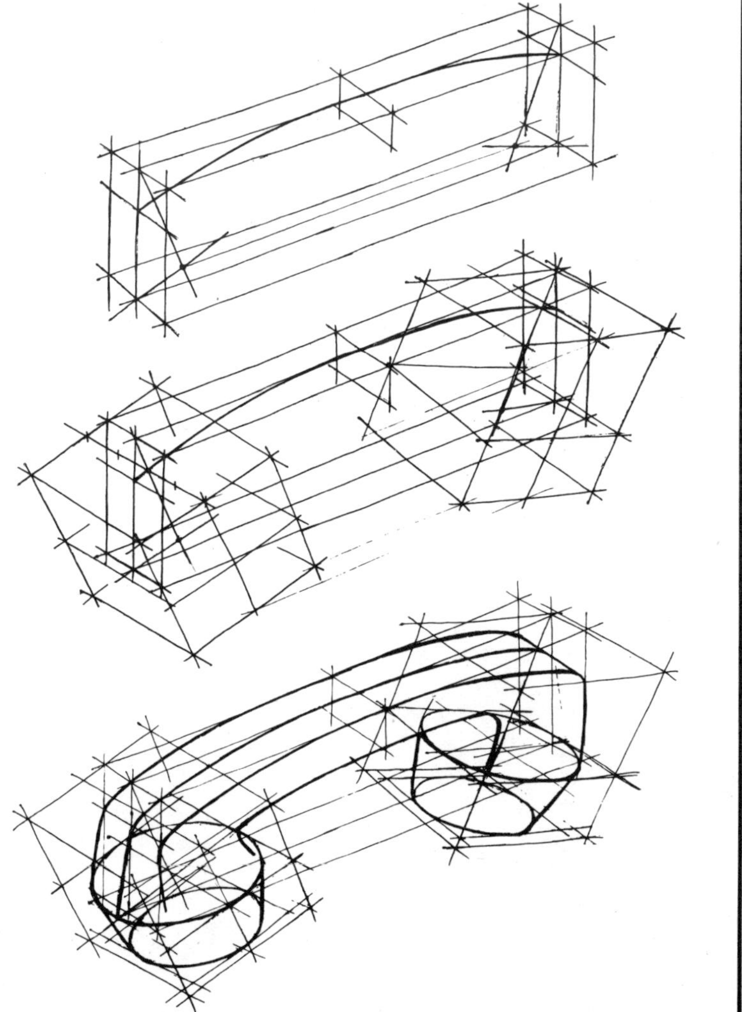

Basic Form Method

In sketching objects where the basic form is approximately spherical, the positioning and proportioning of the adjoining details can be simplified by using auxiliary planes. The circles for the intersecting planes of the sphere are drawn as shown.

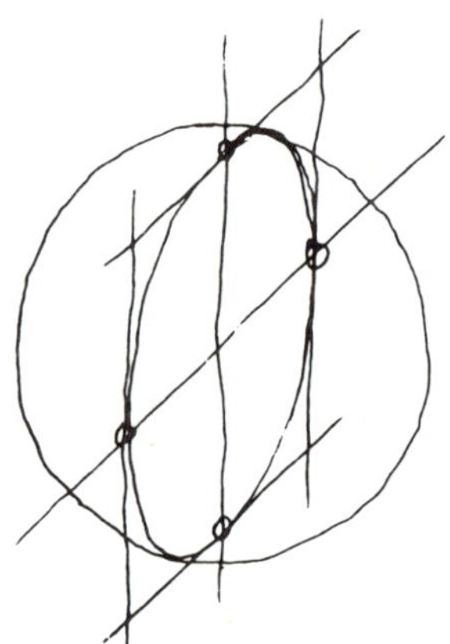

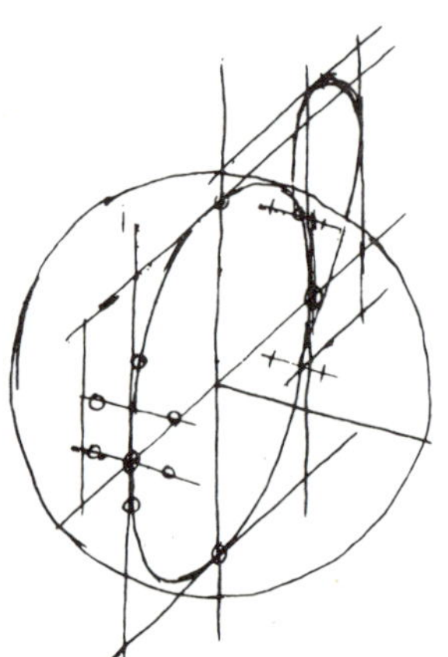

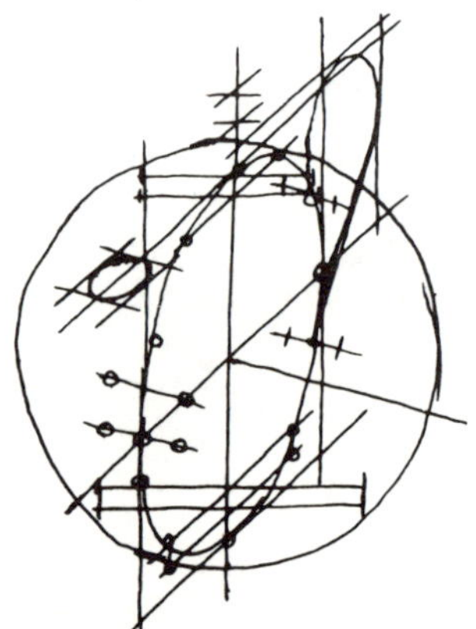

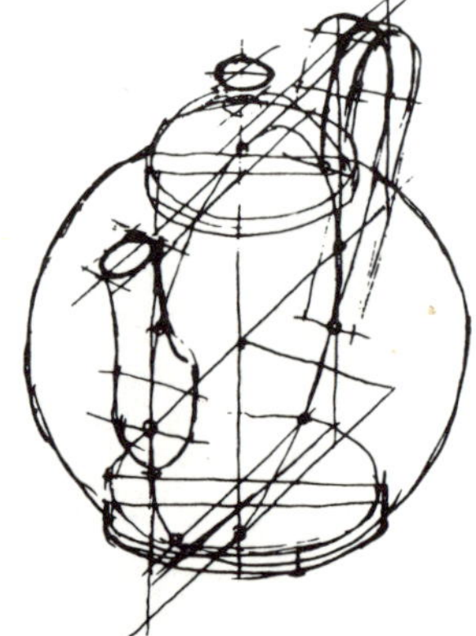

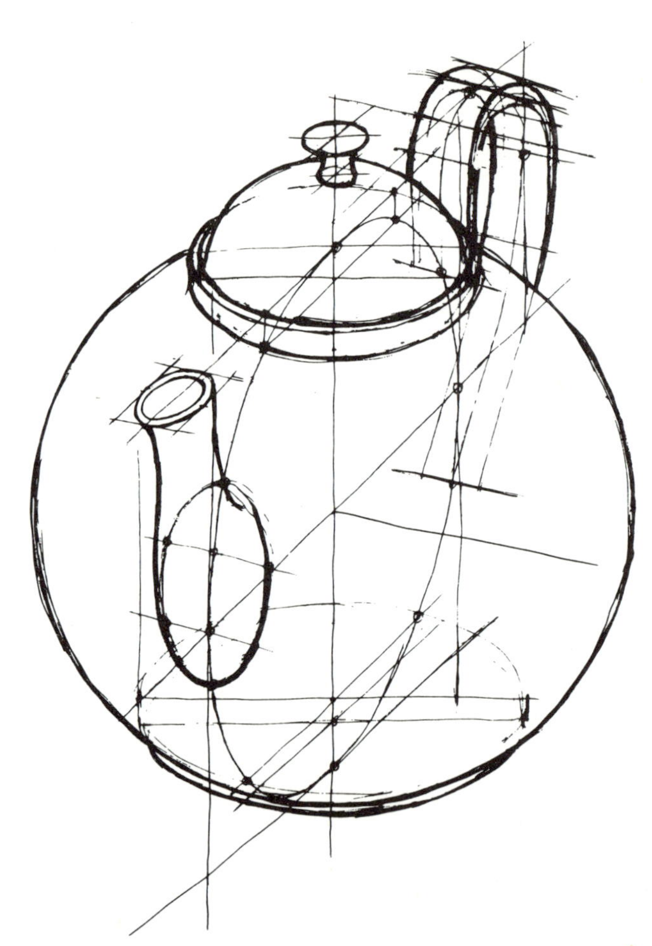

Axis Method

Where objects have components lying on pronounced axes, it is appropriate to start with these axes, and hence decide the type of projection, the positioning of the view and the main proportions. The material around the axes is added afterwards.

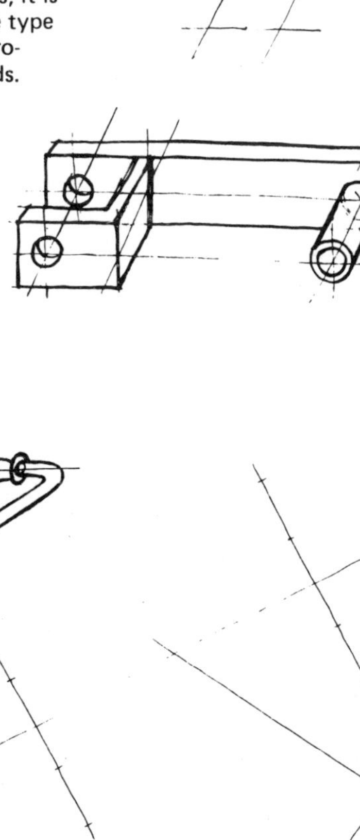

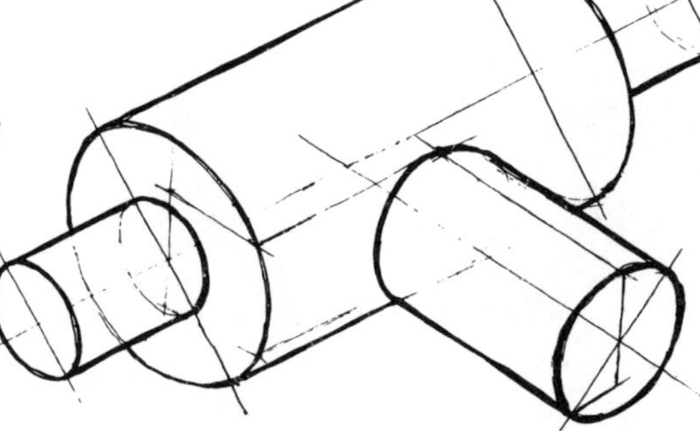

Shading and Contours

It is often useful to use shading and contours to model the desired property, particularly when sketching form. See also pages 35 and 111.

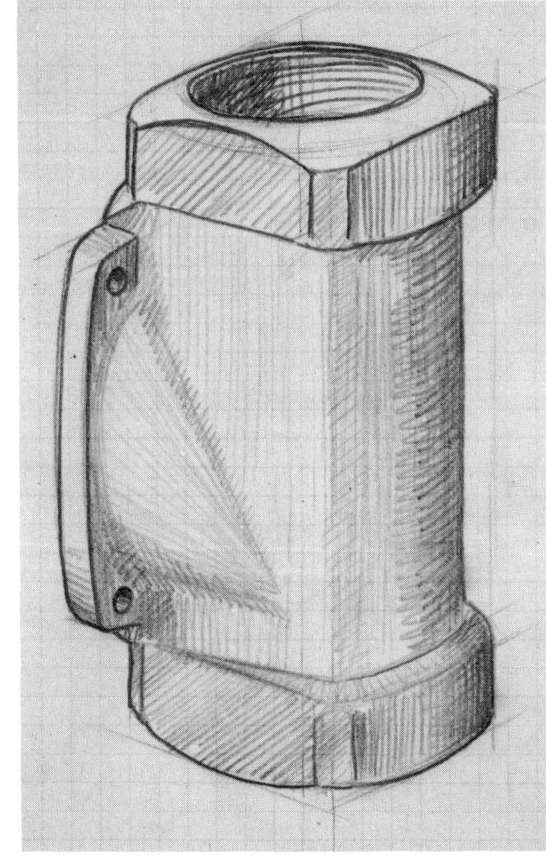

Issigonis's sketch of a car structure combining different techniques, all serving to make the sketch clear, and recording the decisions taken in the phases of design. Note his use of profile sections. The sketch of the valve housing contains many shadows, giving prominence to the soft curved shapes.

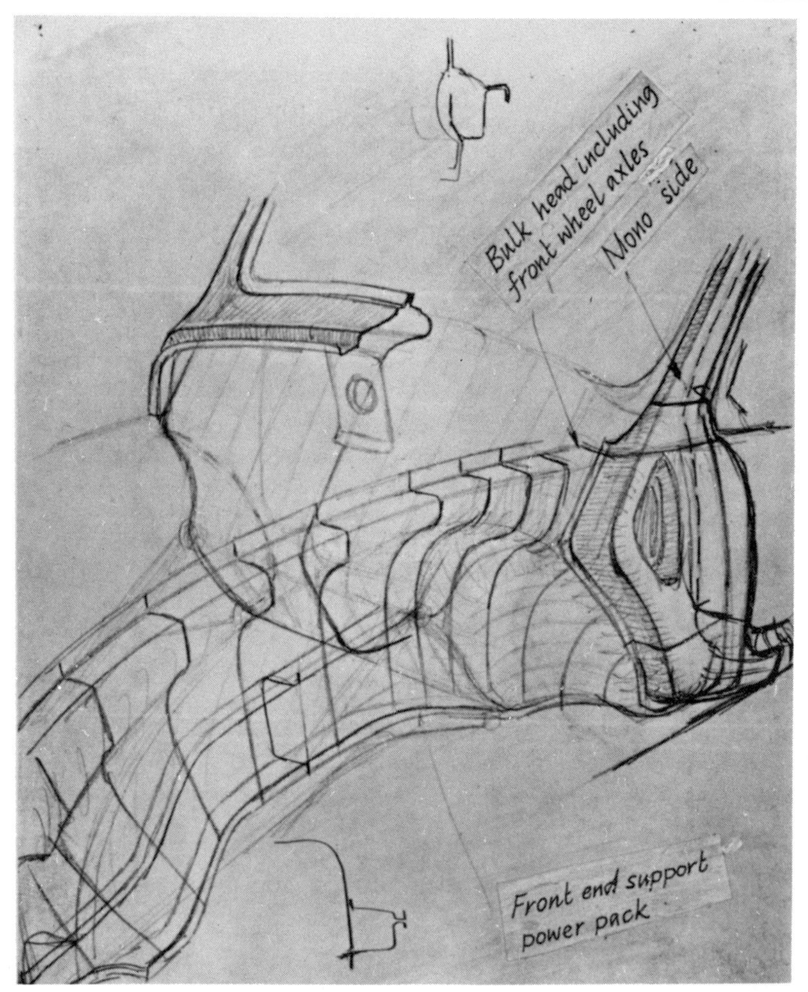

(Reproduced by courtesy of Sir Alec Issigonis and Austin Morris)

Sub-Dividing

In sketching, lines and surfaces often need to be divided in known proportions. The methods shown are simple and an improvement on scaling by eye.

HALVING

The intersections of the diagonals decide the centre:

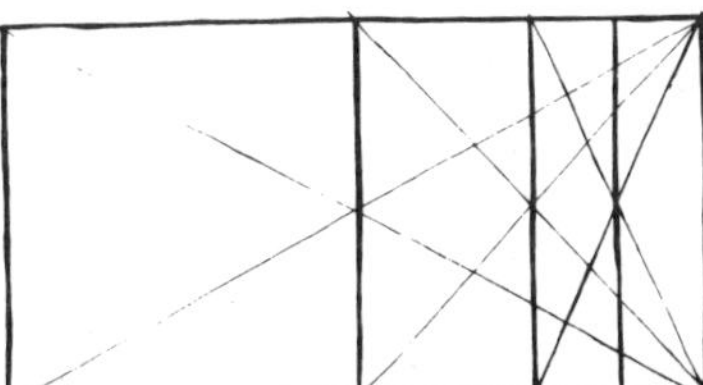

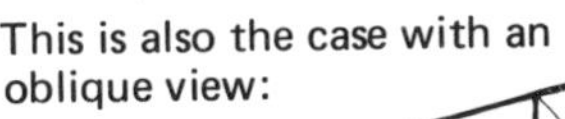

This is also the case with an oblique view:

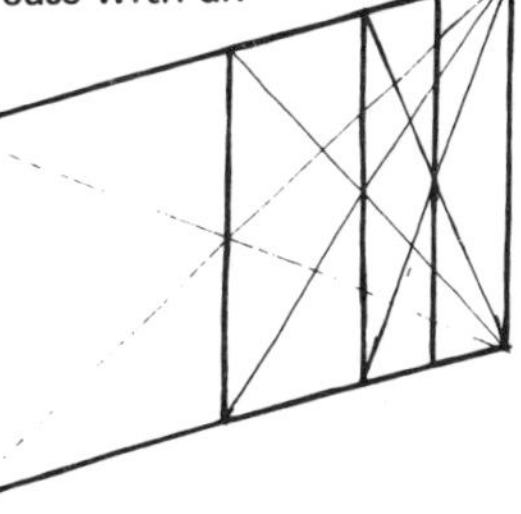

And a view in perspective:

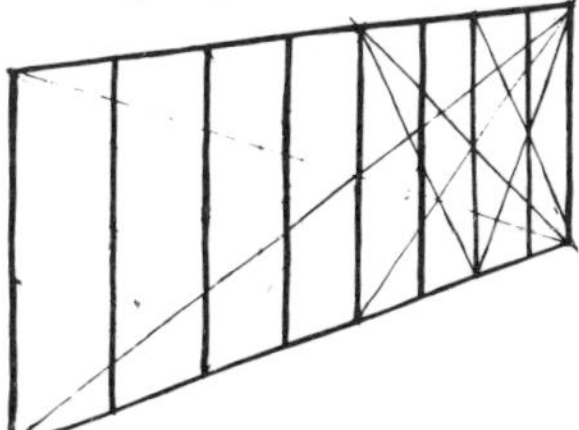

DIVISION INTO EQUAL PARTS

Arbitrary lengths are stepped off along a construction line at an angle to the divided direction and are then transferred as shown in this example of three equal divisions.

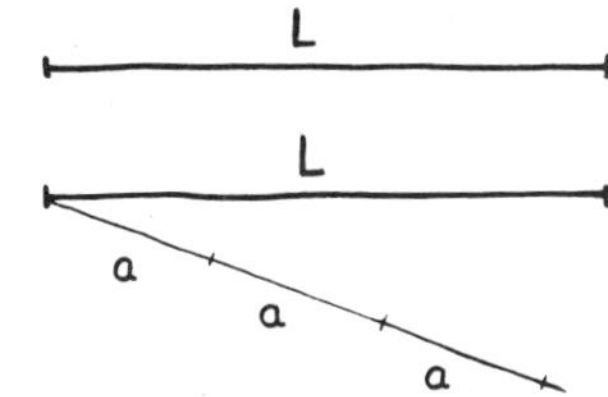

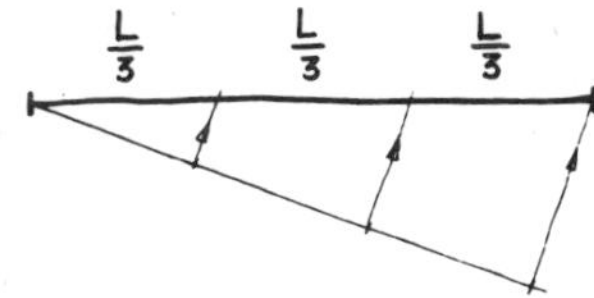

TRANSFER OF KNOWN DIVISION

By using a diagonal construction line, a division in a known direction may be transferred to divide a line in another direction.

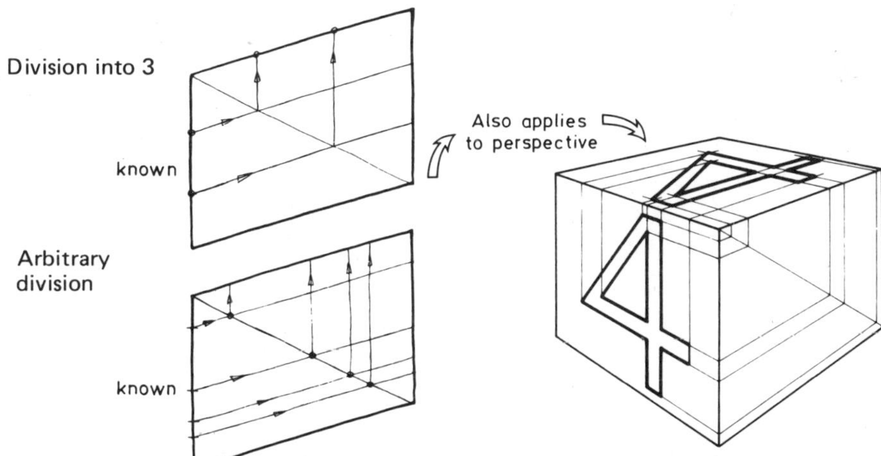

Sketching Circles

Circles are drawn as ellipses in pictorial projections. In isometric and similar projections, the major axis is at right angles to the normal of the plane containing the "circle". This is not the case in oblique projection.

The sides of a parallelogram enclosing an ellipse become tangents to the ellipse, and the mid-points of the sides are the touching points. These four points of tangency are usually sufficient to enable an ellipse to be sketched, but the major axis may also be drawn.

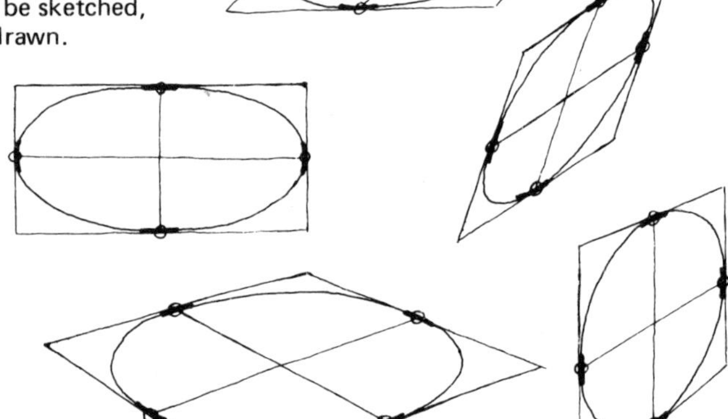

Construction lines where additional supporting points are required

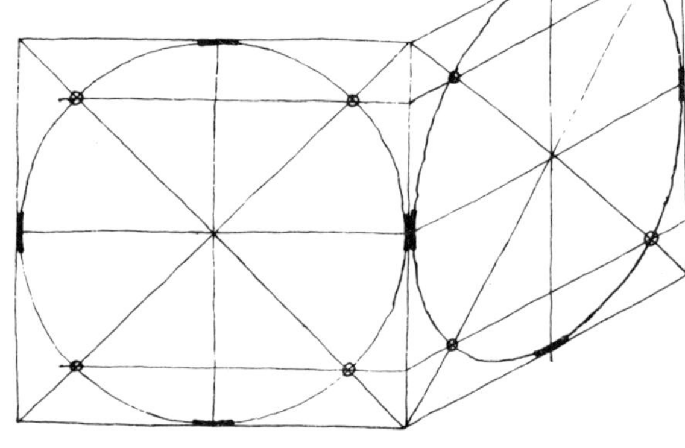

Construction of an ellipse

Various constructions for the points of tangency and additional supporting points, through which an ellipse can be drawn are shown. An approximate ellipse can also be constructed using arcs of circles.

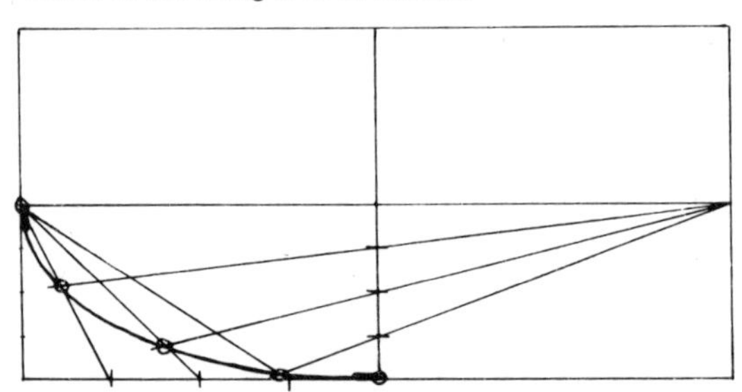

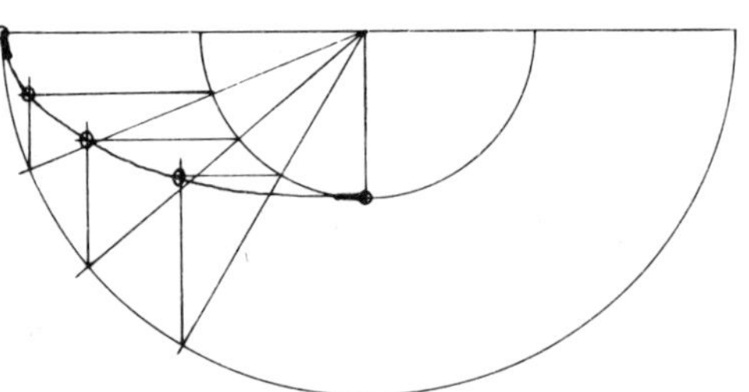

Determining points on an ellipse

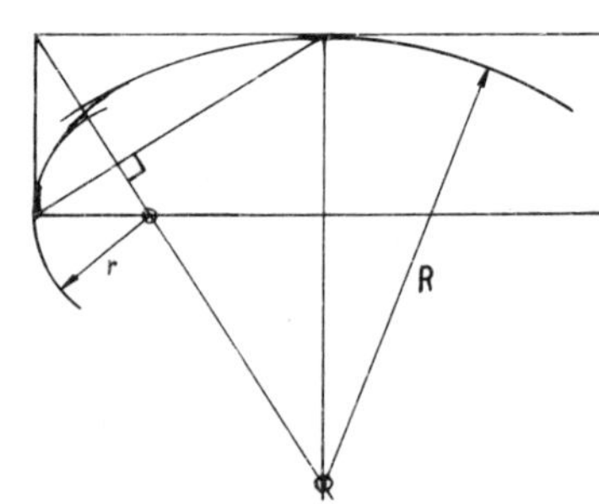

Determination of circular arcs

Intersection of Planes

When two planes intersect, if two points on the intersection are known, then the line of intersection can be drawn.

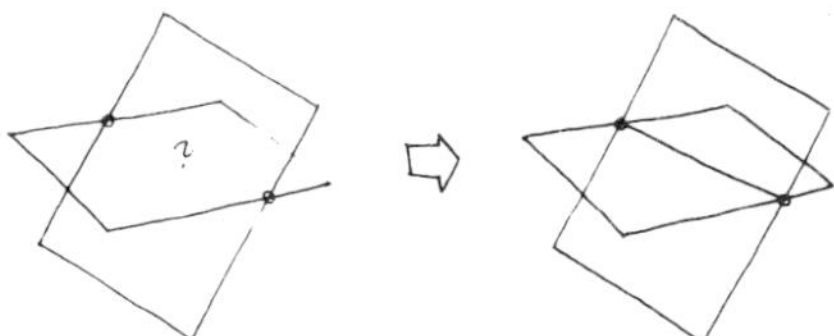

EXAMPLE: Intersection between a prism and an inclined plane

The prism is defined by its end ABC, lying horizontally in a plane which intersects the inclined plane along the line (l). The prism sides are vertical. The position of an apex e.g. A, relative to the line l must be known.

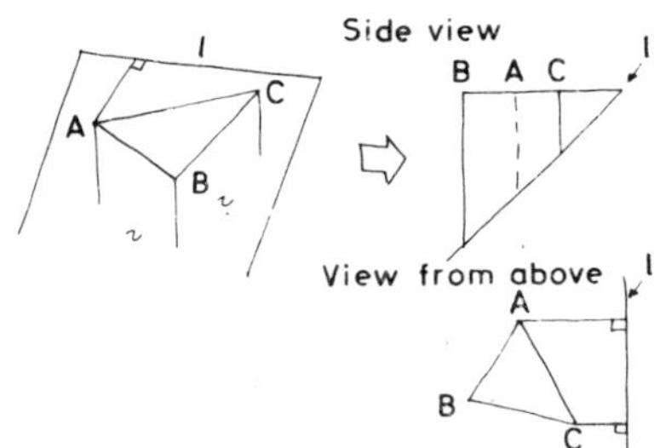

METHOD

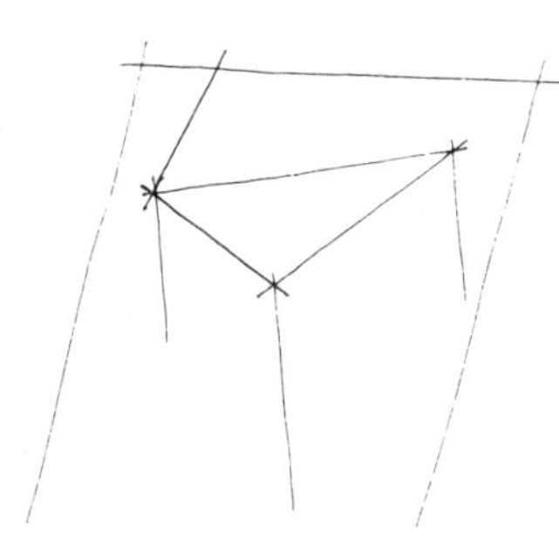

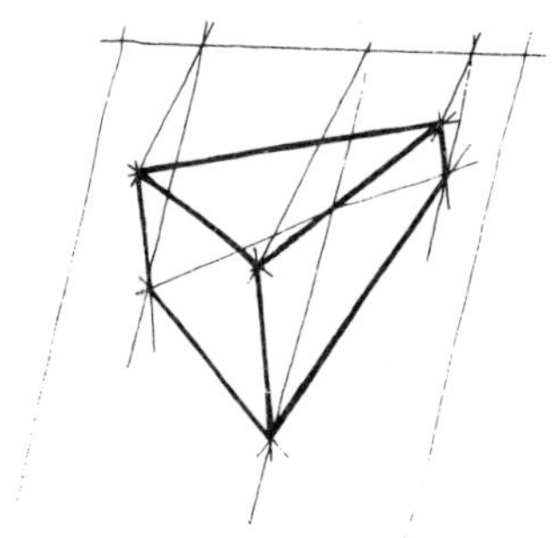

EXPLANATION

The prism, the inclined plane, and the line A_1A are drawn; this line is pictorially at right angles to line (l).

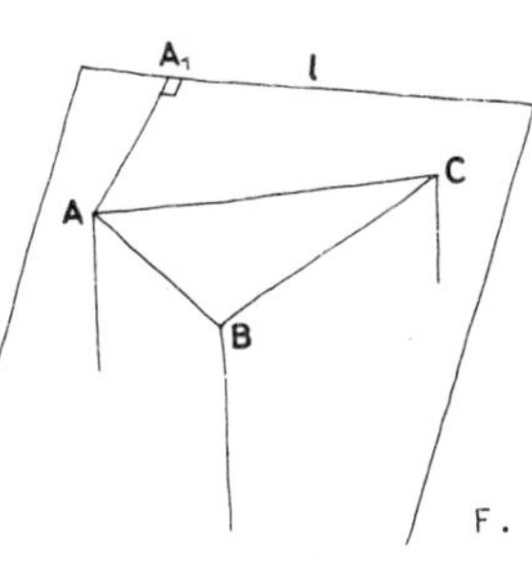

A line is then drawn on the inclined plane from A_1, and pictorially at right angles to the line (l), its intersection with the vertical edge of the prism gives point D. The method is repeated for E and F.

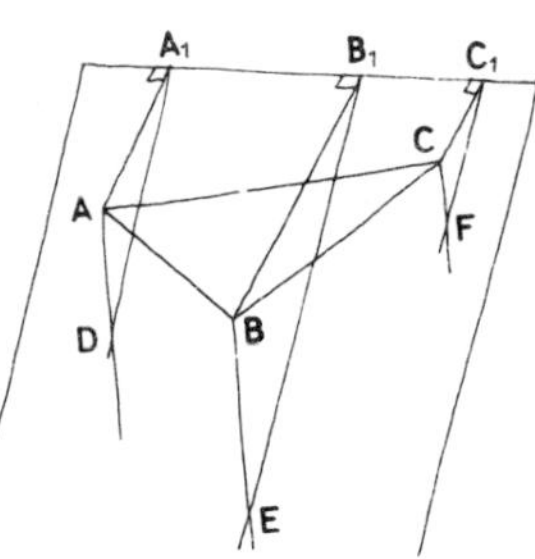

The lines of intersection between the faces of the prism and the plane are drawn by connecting D, E and F.

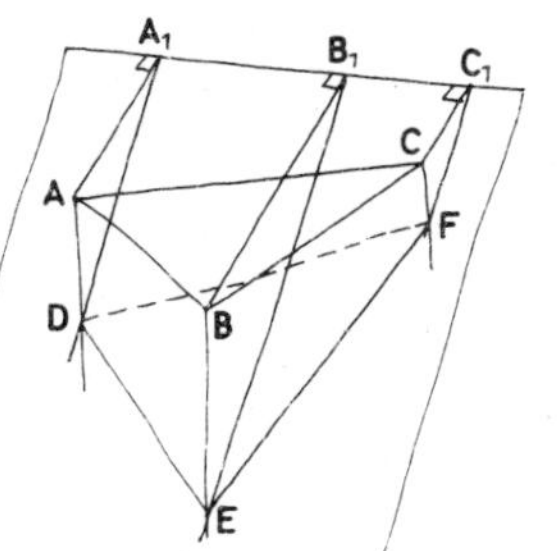

Intersection Between an Inclined Plane and a Cylinder

METHOD

EXPLANATION

The relative position of the plane is known by defining a horizontal plane containing ABC and its line of intersection (ℓ) with the inclined plane.

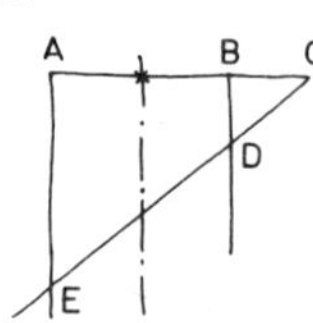

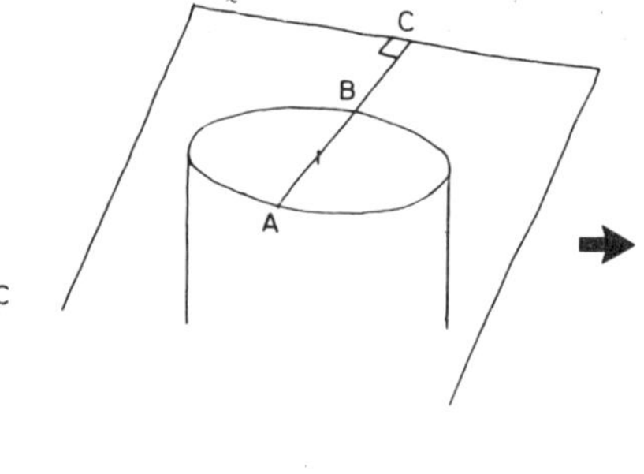

A construction plane is established having the following properties. It contains the axis of the cylinder, it is pictorially at right angles to the inclined plane and pictorially at right angles to the line (ℓ). The intersection between the inclined plane and the construction plane gives the line through CD and E, lines can be drawn vertically through A and B to give E and D. The tangents at these points are parellel to (ℓ).

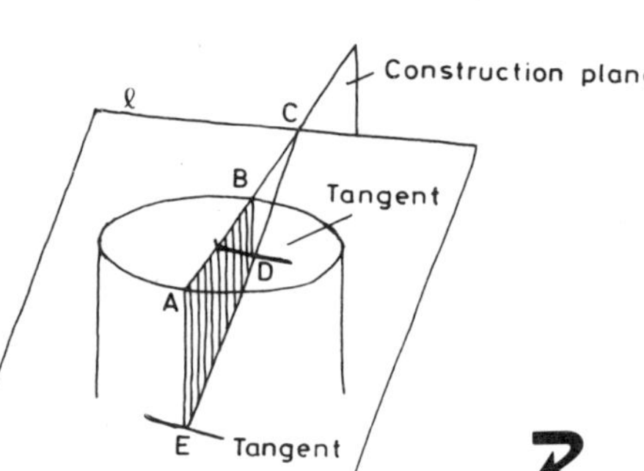

Other points on the curve of intersection can be found by establishing further construction planes parallel with the first. It is only necessary to establish points for one half of the intersection because the intersection on the other side of the central construction plane is a mirror image i.e. G and F are the same distance from the construction plane as G′ and F′.

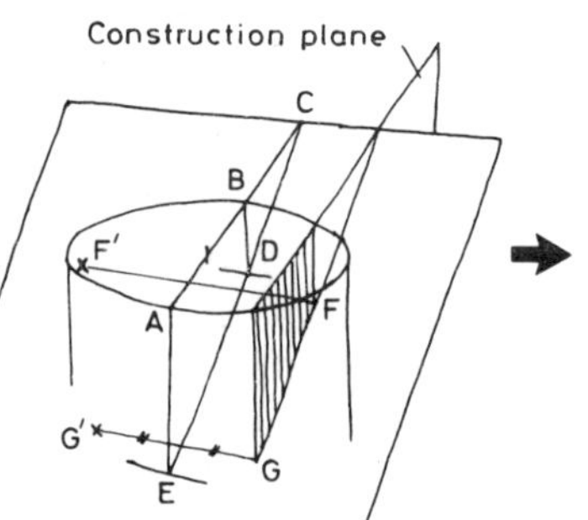

The position of the last construction plane is selected to be tangential to the cylinder, as a consequence the tangents at H and H′ are parellel with DE.

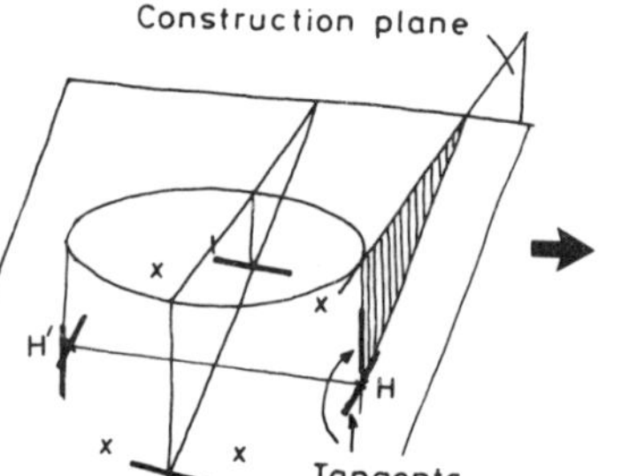

The complete curve of intersection can now be drawn. Note the usefulness of knowing the positions and directions of the four tangents through which the circle is drawn.

Curves of Intersection

Intersection of curves and an inclined plane

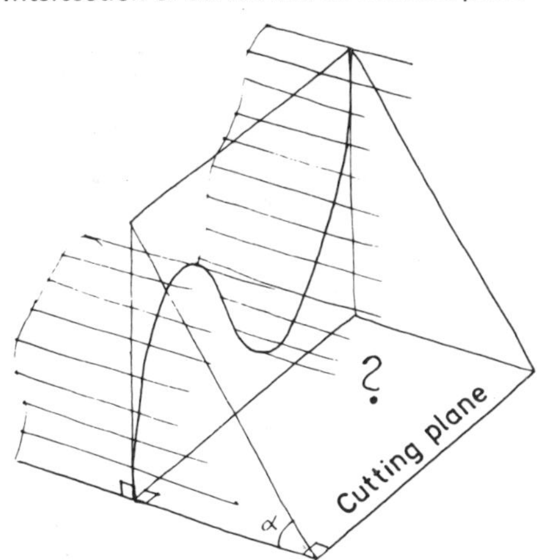

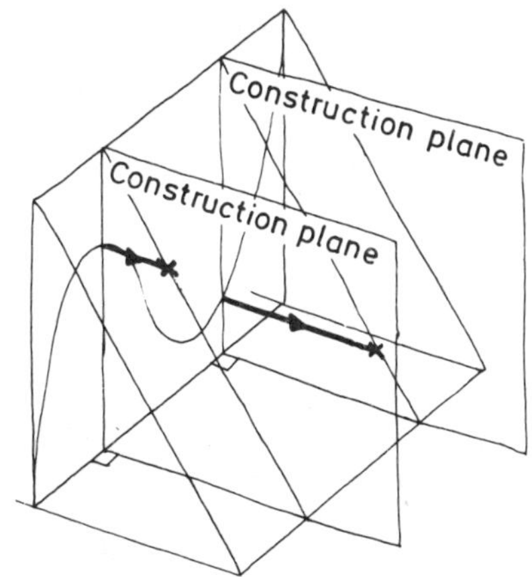

Determining the construction points

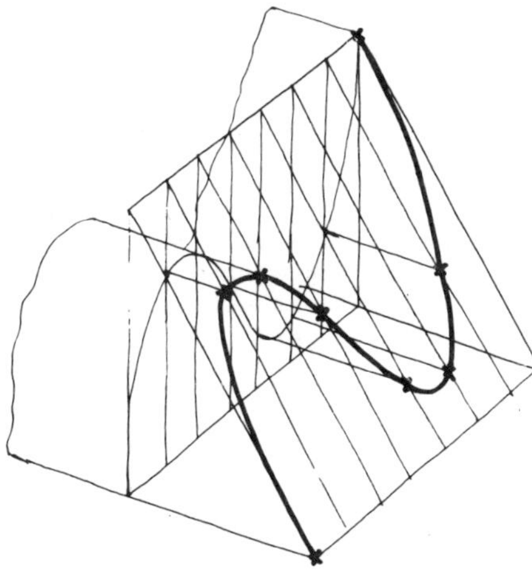

The intersection curve

Intersection between two cylinders:–

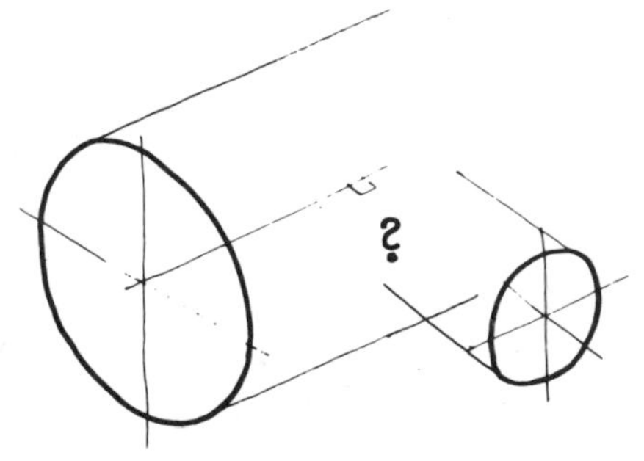

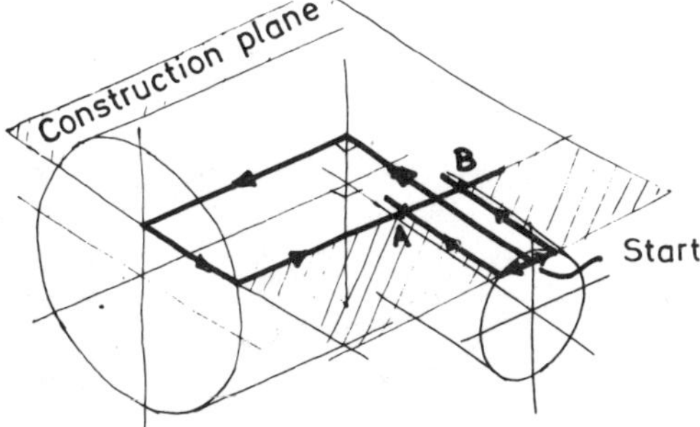

The points are decided in the order given by the arrows. A and B are points on the intersection curve.

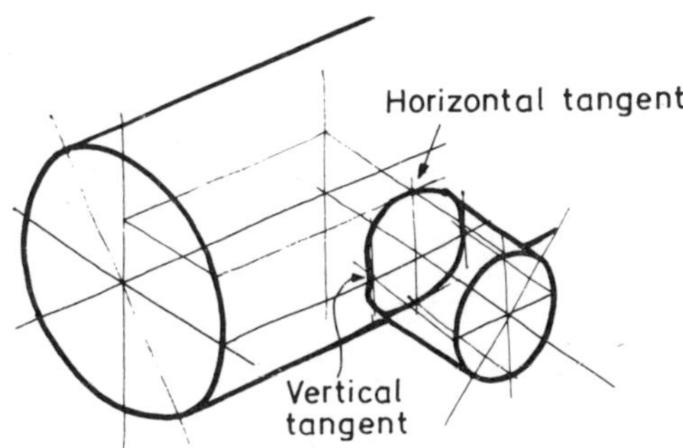

The intersection curve

4.2 DRAWING TECHNIQUE

BOLD FREEHAND DRAWING

Use

Bold freehand drawings are used where there is a need for clear simple drawings, e.g. for printed instructions, lectures and illustrated articles.

Aids

Bold drawings are those where important lines are given prominence and all construction lines are removed. This can be done by lining in and rubbing out or by photographic techniques. The main aids are drawing tools with thick black points and possibly photographic equipment.

Methods

The boldness can be created in the following ways:

A. The figure is drawn in simple heavy clear lines. Squared paper may be used.
B. Feint construction lines are retained, whilst the contours are emphasised in heavy black lines, possibly with another kind of drawing tool (drawing pen, felt tip pen). This technique is suitable for flip charts.
C. Construction lines can be removed by a rubber, perhaps by lining in, rubbing out and lining in again. Acceptable illustrations can be produced in this way.
D. Lining in can be done on a new sheet, e.g. by tracing or by projecting the original using an epidiascope. Semi opaque paper or tracing paper are suitable for tracing, but ordinary paper can be used if an illuminated tracing table is available. Tracing allows the use of sketches at different stages in their development and alternatives can be shown from the basic concept.
E. Photographic reduction can remove greyness and sometimes mistakes and uncertainties in the lines. The result is improved quality. It is used to advantage for illustrations, slides, overhead projector transparencies, etc.

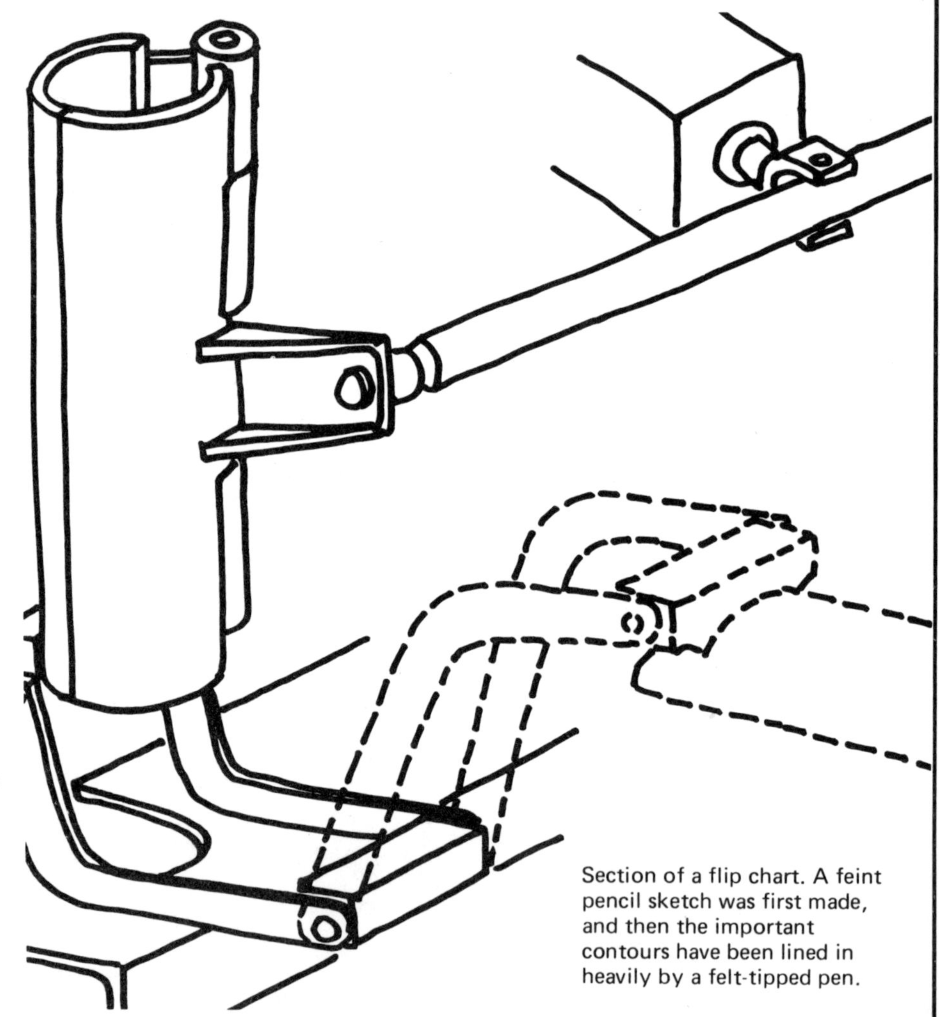

Section of a flip chart. A feint pencil sketch was first made, and then the important contours have been lined in heavily by a felt-tipped pen.

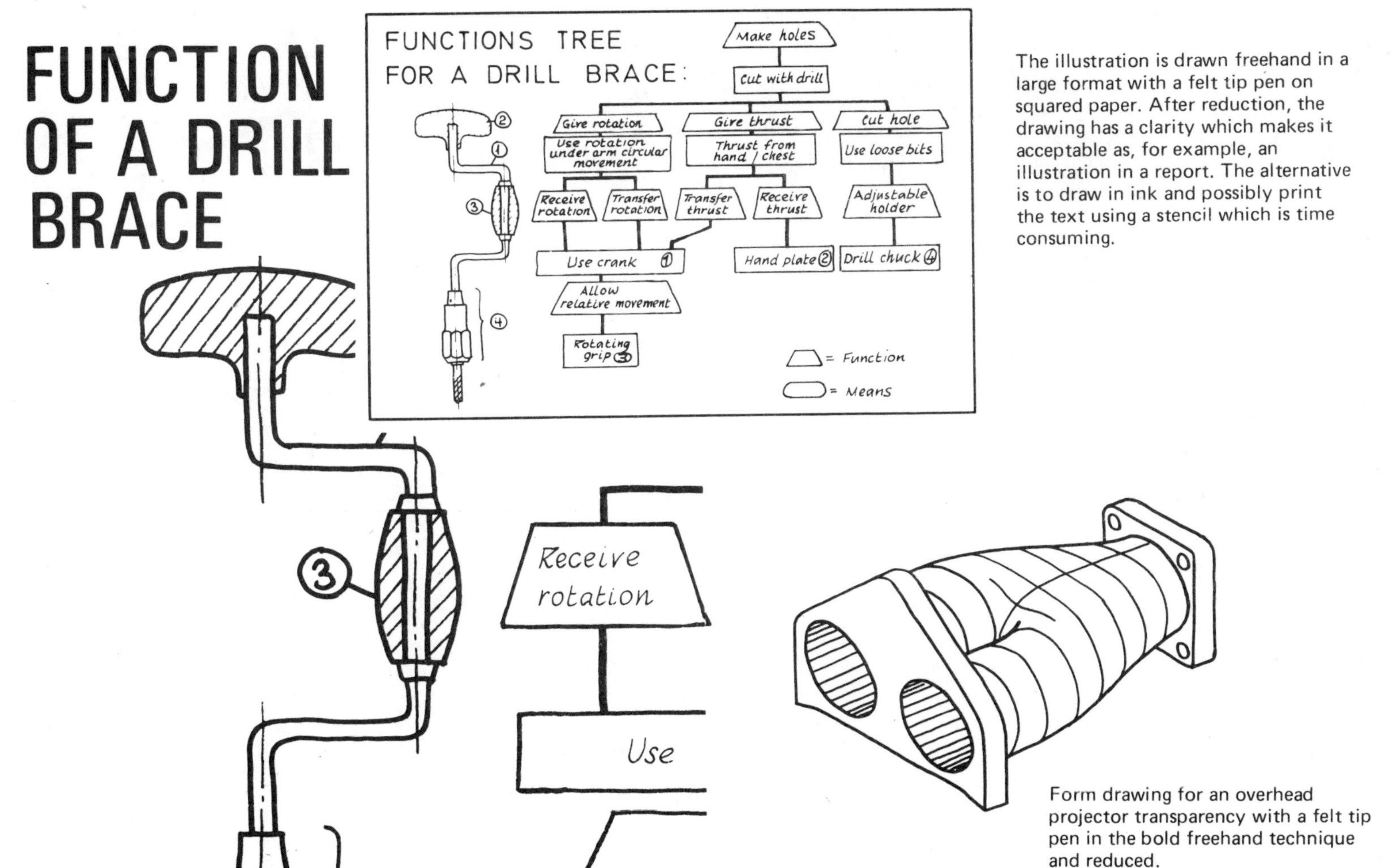

The illustration is drawn freehand in a large format with a felt tip pen on squared paper. After reduction, the drawing has a clarity which makes it acceptable as, for example, an illustration in a report. The alternative is to draw in ink and possibly print the text using a stencil which is time consuming.

Form drawing for an overhead projector transparency with a felt tip pen in the bold freehand technique and reduced.

Examples

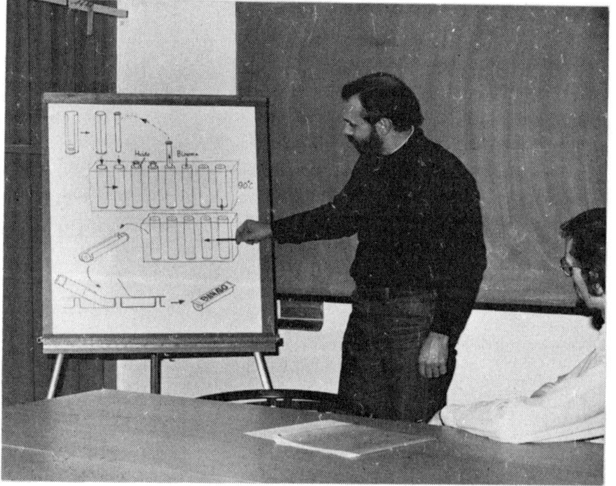

Flip chart with drawings in heavy black lines. The drawings have been outlined in pencil and lined in with a felt tip pen. Lining in can be conveniently performed whilst presenting the drawing.

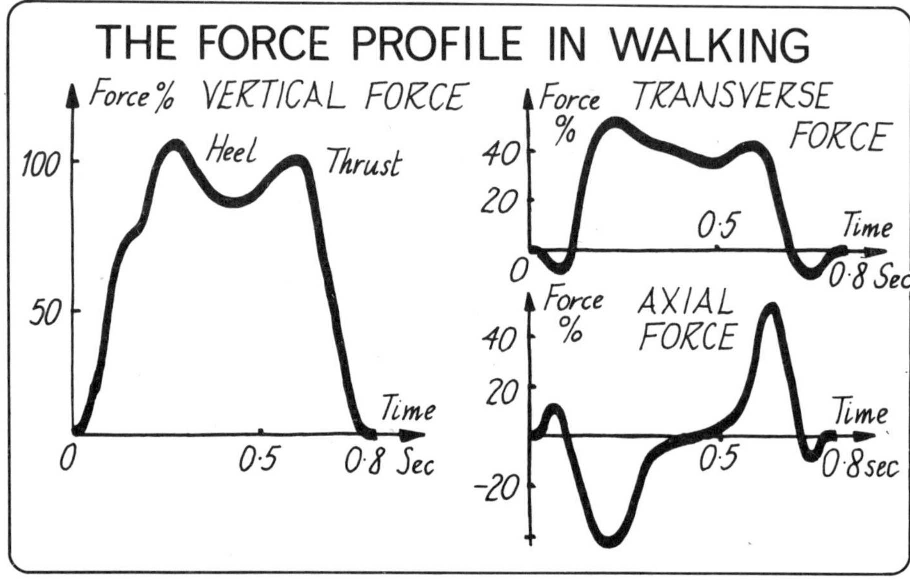

Reproduction of a slide made by photographing a completed freehand A3 sized original. Note the minimal effort of including a few details and notes. The height of the letters should be 2–3% of the frame size to be readable.

The drawing of people is assumed to be difficult. The result can be successful if the form is simplified and drawn boldly.

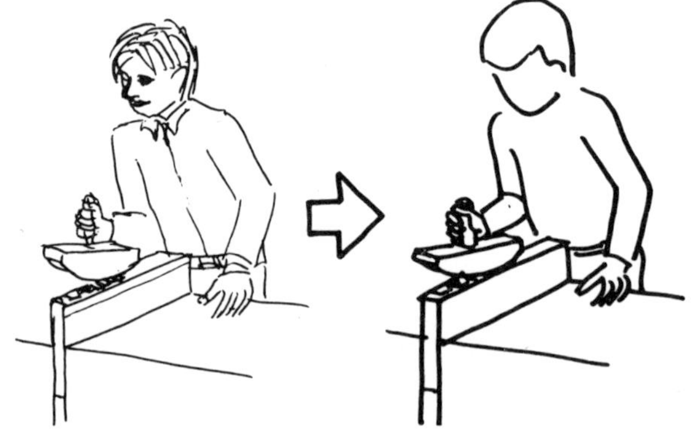

DRAWING WITH A STRAIGHTEDGE

Use

Straight edges and setsquares are often used to obtain straight lines. The straightedge gives exactness, and, when used in geometrical construction with a scale and compasses enables dimensions to be decided. Correctly proportioned drawings can also be produced, e.g. minor detail drawings and perspective drawings.

Technique

The straight edge still permits a differentiation of the line enabling construction lines to be faint. Clear precise drawings are generally achieved with a straight edge.

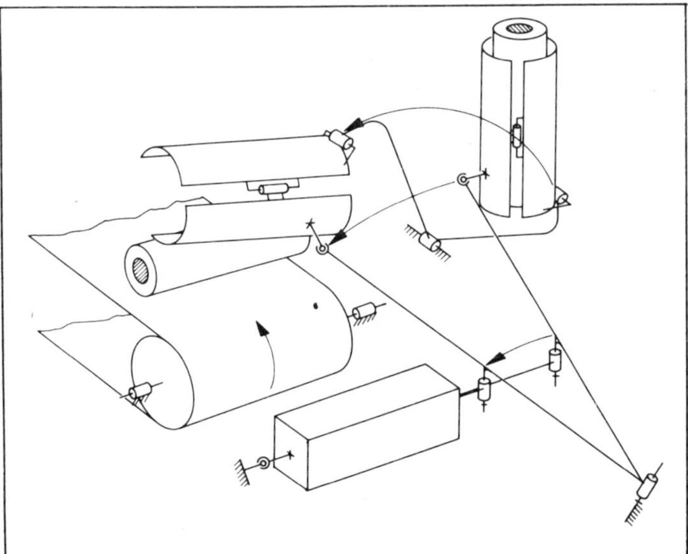

This illustration was produced in three phases: (i) a freehand pencil drawing in large format, (ii) traced in ink using compasses, template and straight edge, and (iii) photographically reduced.

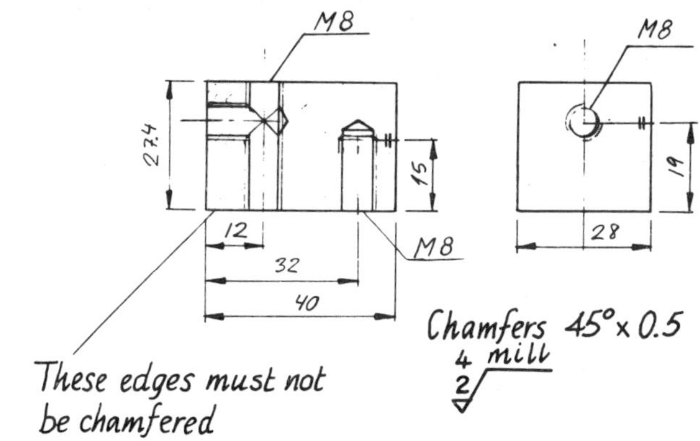

Common situation: Detail drawing with pencil and rule on squared paper.

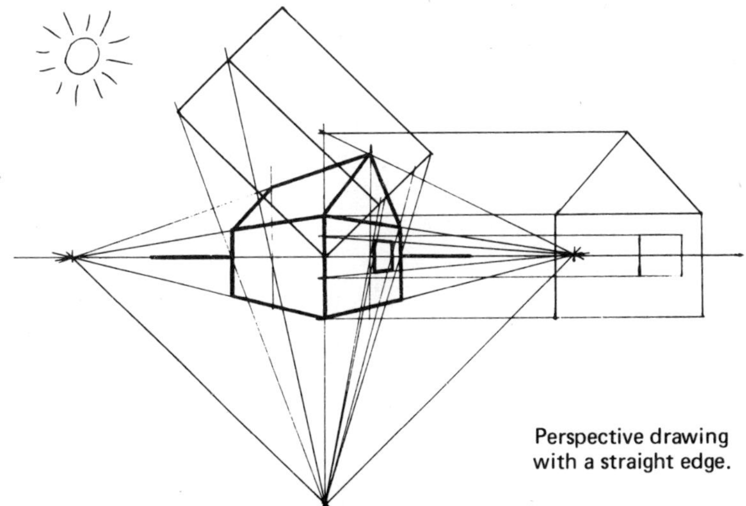

Perspective drawing with a straight edge.

4.4 DRAWING TECHNIQUE

DRAWING WITH A DRAUGHTING MACHINE

Use

Draughting machines are generally used for engineering drawing, i.e. detail and assembly drawings, layouts, and for some illustrations.

Aids

Draughting machines are built according to two principles: parallelogram (and one of two forms) of parallel motion. They may be equipped with refinements such as protractor mounted scales, lockable in a given position, and sometimes having special edges for ink drawing.

TECHNIQUE OF DRAWING LINES

The most important tools are the pencil and drawing pen.

A hard sharp pencil is used to lay out the views, e.g. 2H–4H.

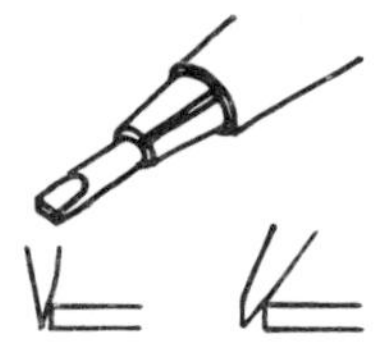

A softer sharp chisel point pencil is suitable for lining in. (Sharpen on abrasive paper). Wide lines can be achieved by making two strokes and inclining the pencil on the second.

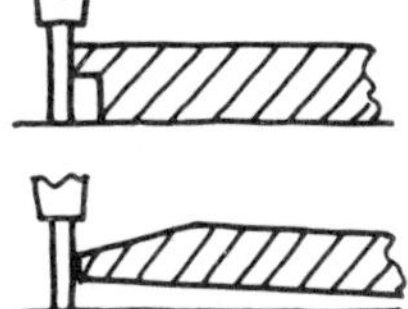

Stylus pens are suitable for lining in with ink. Remember the guiding edge, it can smudge the ink.

REMEMBER: Straight edges are flexible. The lines become curved if too much force is used.

Always hold the left hand on the draughting head and lift the head clear of the paper when repositioning, to avoid dragging the drawing.

PHASES OF DRAWING

Drawing in ink

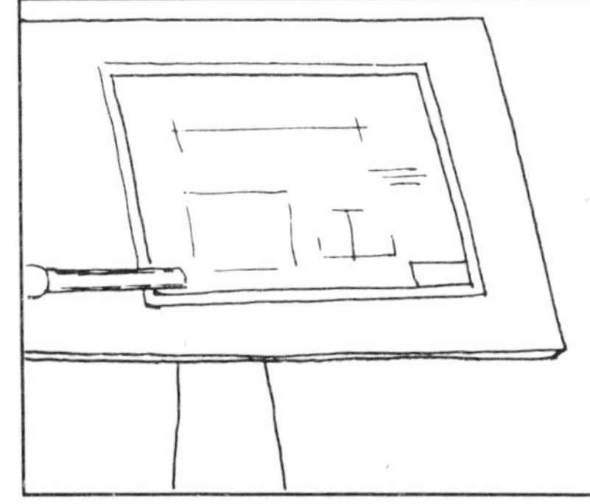

1. Plan the drawing so that the number of views and their positions are clear.

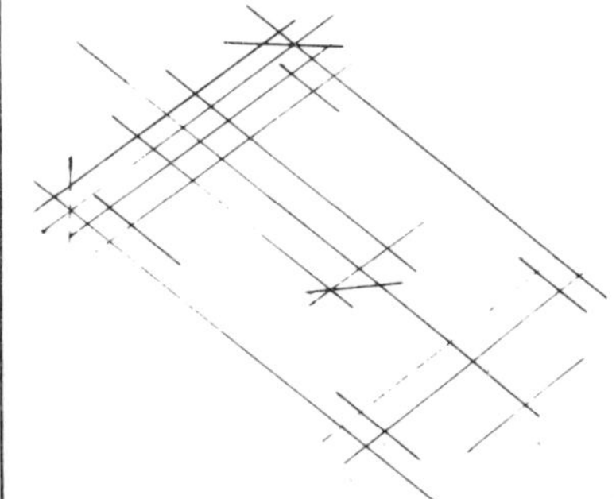

2. Make a preliminary drawing in feint lines, remembering that it is to be lined in later, i.e., hatching etc., is omitted and the lines tend to be 'long'.

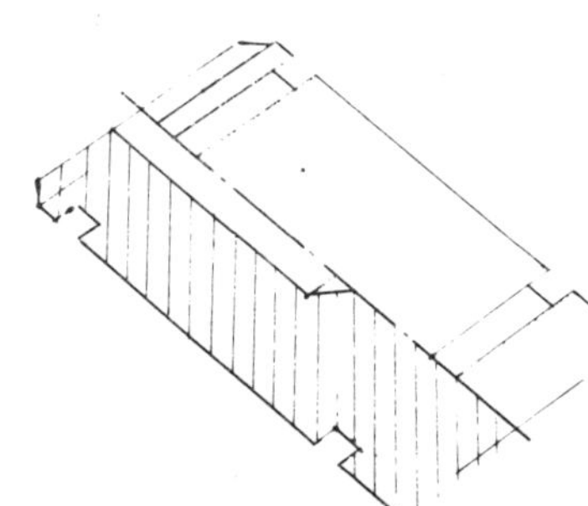

3. This drawing is too detailed, and the details have been drawn too carefully.

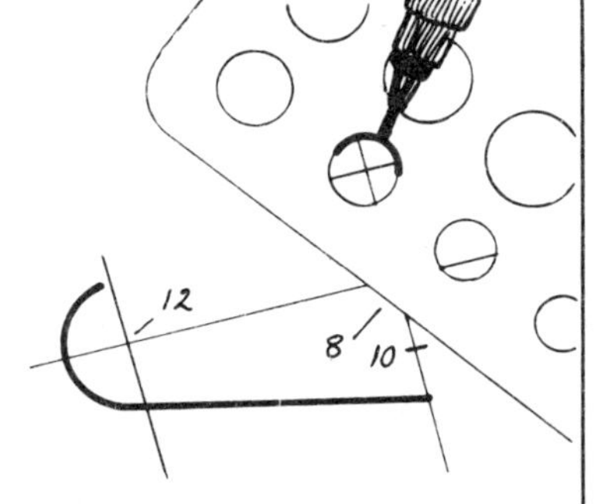

4. The drawing need not be complete before lining in. Circles can, for example, be left out.

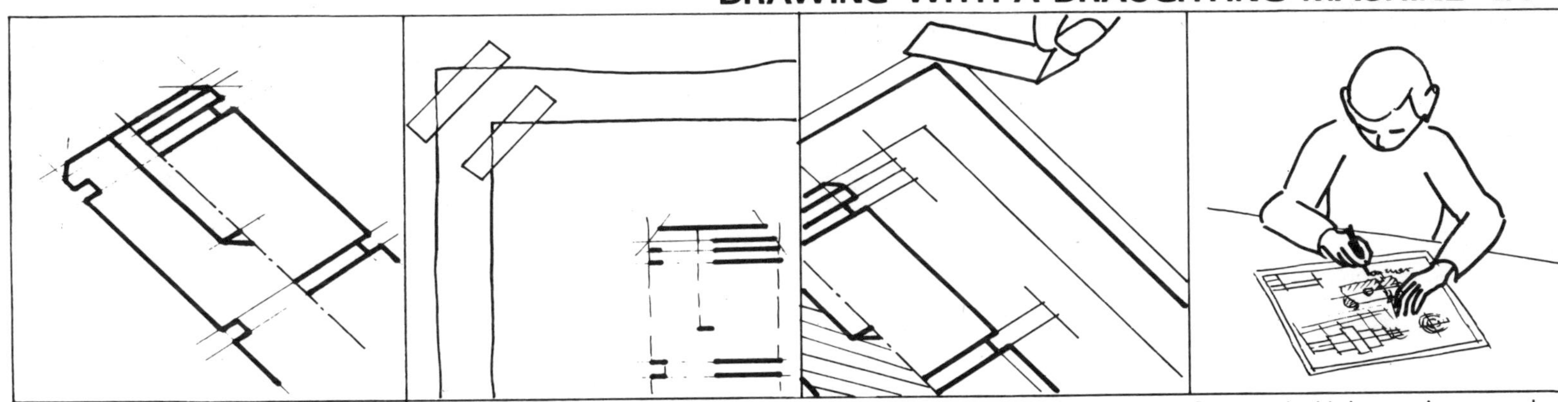

5. Drawing conventions must be observed when lining in.

6. Lining in on a second sheet is wasteful and introduces errors.

7. It is more convenient to remove the drawing, . . .

8. . . . and add the text in a normal writing position.

PHASES OF DRAWING Drawing in pencil

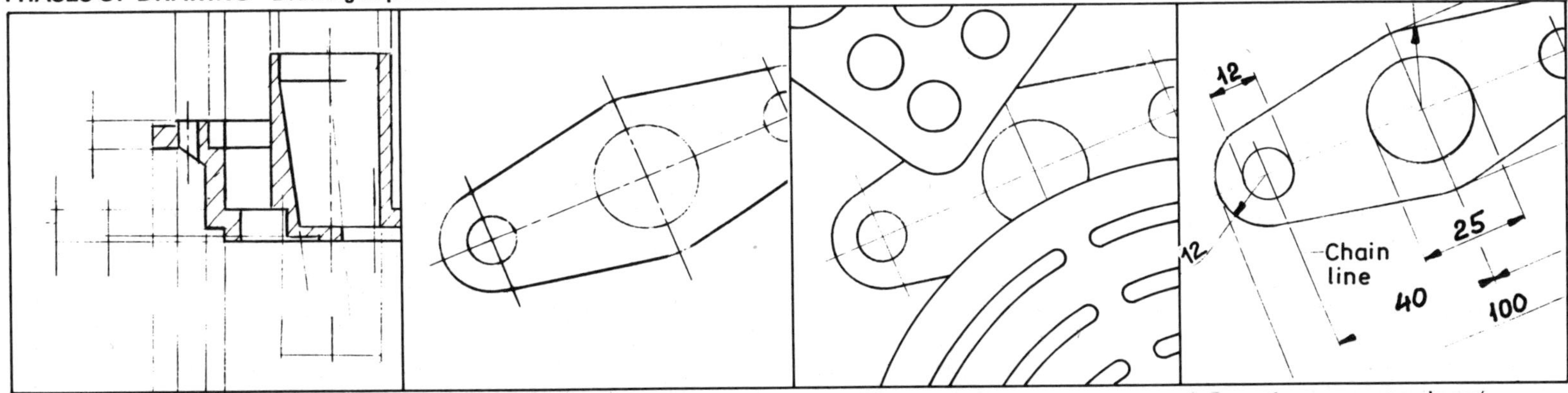

1. Make a preliminary drawing with a sharp hard pencil, line in with a softer chisel pointed pencil. If the line is powdery the pencil is too soft, if the line is grey it is too hard or blunt.

2. It may be difficult to get circles really black with a pencil.

3. Use templates instead.

4. Sometimes arrow heads and dimensions are in ink for clarity.

4.5	DRAWING TECHNIQUE

USE OF TEMPLATES

Use

Templates cover a wide spectrum from ship curves to alpha numeric stencils.

Aids

Ship curves, French curves, Circle templates, Radius templates, Ellipse templates	Relevant for all kinds of drawings with accurate lines
Lettering stencils	Many types of lettering for anotating drawings and illustrations
Screw/nut templates, Weld templates, Circle templates (small internal)	For drawing details.
Diagram templates Computer flow charts, Pneumatic, Hydraulic, Electronic, Heating, water sanitary installations Electrical installations	Representation of details on specialised drawings can, with advantage, be drawn with a template.
Variety of specialist templates Arrows, Furniture, Valves	See specialist suppliers catalogues.

THE TECHNIQUE OF DRAWING LINES

Two points should be stressed: When lining in ink, the ink can be drawn under the template and smudge. Solutions: →
Circular templates tend to be inaccurate because of the pencil thickness.

Transfers

A closely related area is the use of transferable letters and symbols. Many symbols and details e.g. arrows, circles, shading, are available. Transferable letters are well suited as text for illustrations, exhibition material etc.

PHASES OF DRAWINGS AND METHODS

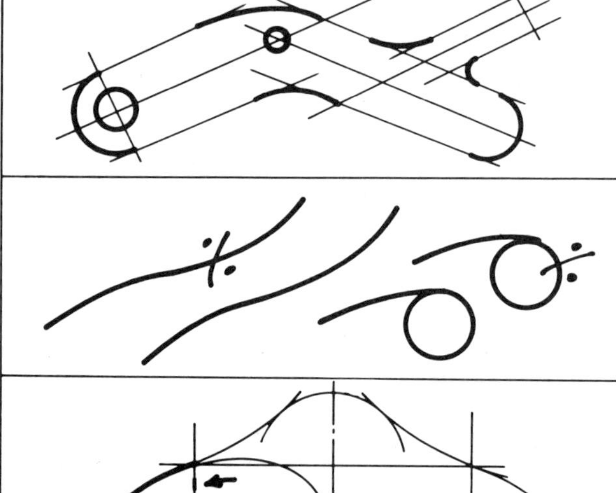

It is easier to make straight lines blend with curved lines, consequently curves and the circles should be drawn first.

Care must be taken in combining curves. It is better to first try the combination lightly in pencil.

With symmetrical curves, it is a help to mark the template before it is turned.

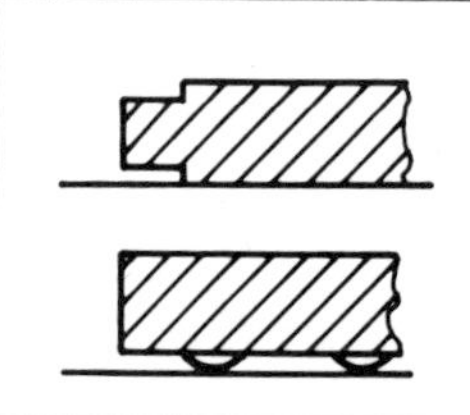

Certain templates are for use with ink, and have a special edge or spacer's so that the template is held clear of the paper.

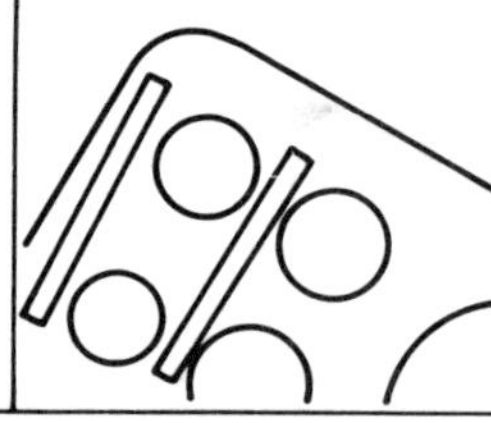

Templates can be improved by sticking tape to the underside, to separate them from the surface

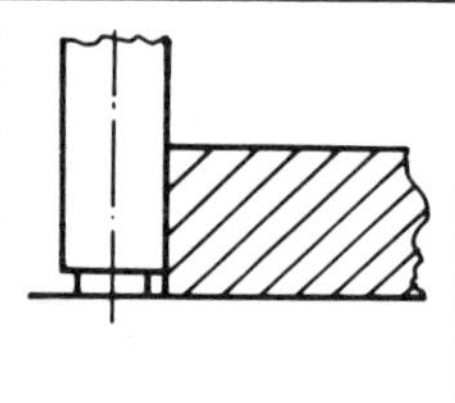

Special pens are made for use with templates e.g. some lettering pens.

DRAWING WITH A PLOTTER

Use

Many computers have the facility to produce drawings on a plotter or screen. On a screen, instantaneous illustrations are available saving time consuming drawing.

It is outside the scope of this book to go into further detail. Some of the possibilities are shown in the examples.

In many cases standard sub-routines are available for generating text, drawing axes, scaling etc, these should be exploited whenever possible.

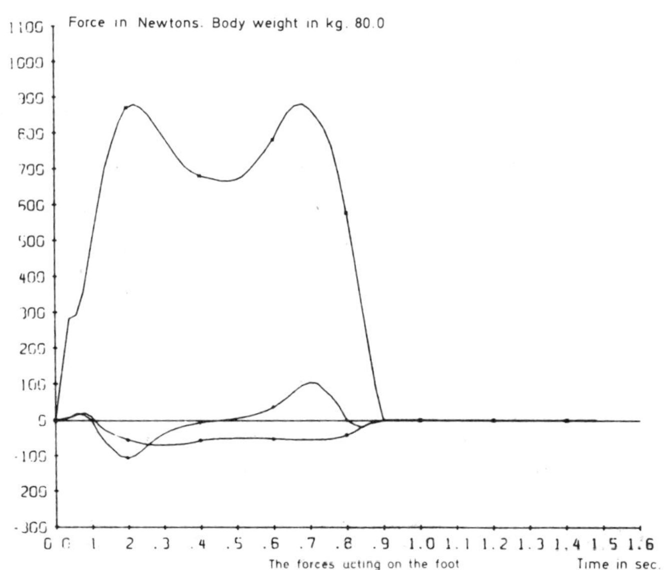

The plot of a curve and the marking and labelling of the axes.

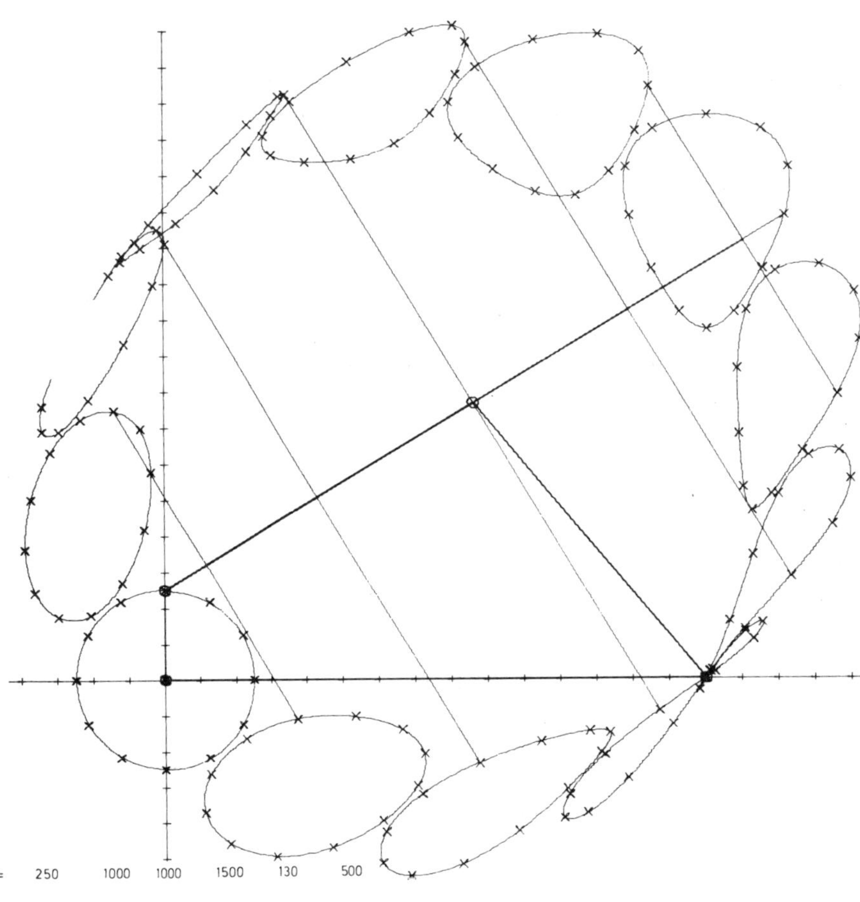

The plotter output for the movement of a mechanism at different positions. The crosses on the curves indicate time intervals.

5. TYPES OF DRAWING

5.0 TYPES OF DRAWING

SUMMARY

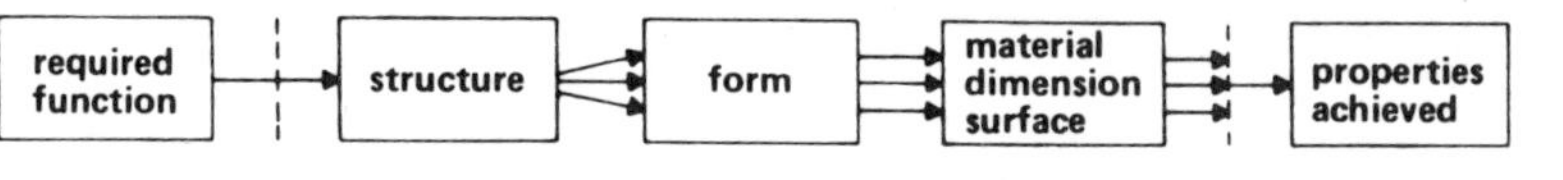

The introduction shows that the type of drawing, the technique and the code depend on the receiver. However, from experience, the designer can usually select the type of drawing for a particular situation.

The following procedure sheets are aimed at the latter. They deal with the more important characteristics of a number of known drawing types. The sheets therefore serve to define the methods and applications for each type of drawing.

The code, but not the technique, is normally specified for every type of drawing. Care should therefore be taken to choose the technique best suited to the given situation.

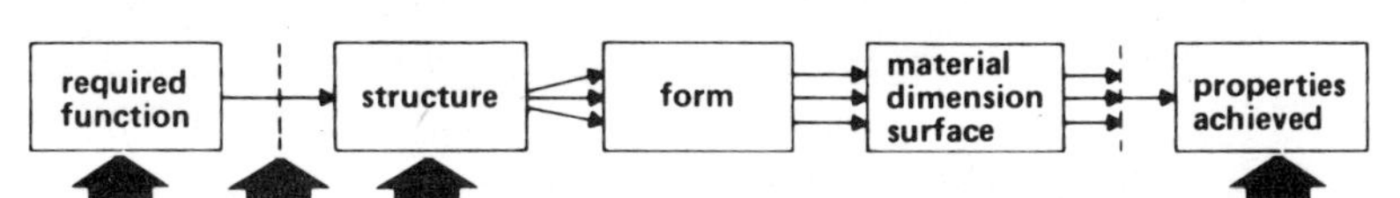

BLOCK DIAGRAMS

Term

Block diagram.

Use

For describing the relationships between elements in a group. These may be elements in a mechanical system or in a process. Block diagrams can be used in a number of situations where systems or processes are to be described at an abstract level.

Characteristics

The block diagram comprises boxes (blocks) and connecting lines.

The boxes: represent elements. Different categories of elements are sometimes drawn according to a specific notation.

Examples of symbols which might be used:–

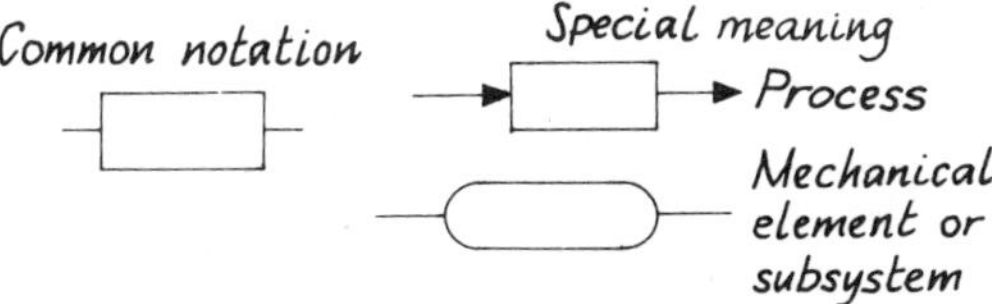

The connecting lines: represent the relationships between the elements. The lines can, if necessary, be given arrows to indicate direction in the relationship. The relationships could be:

- material transfer
- energy transfer
- information transfer
- a physical element at a lower level of analysis

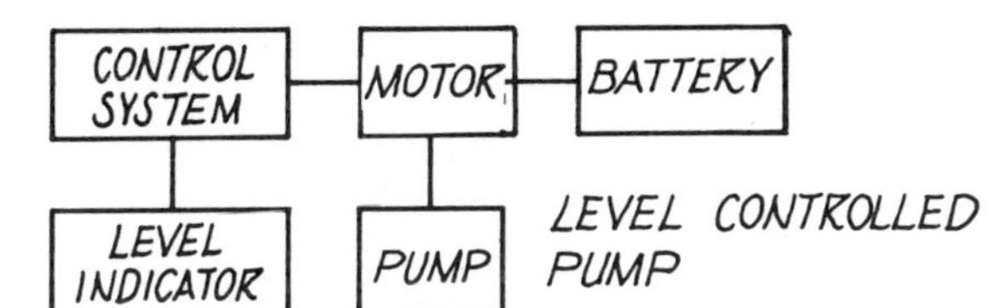

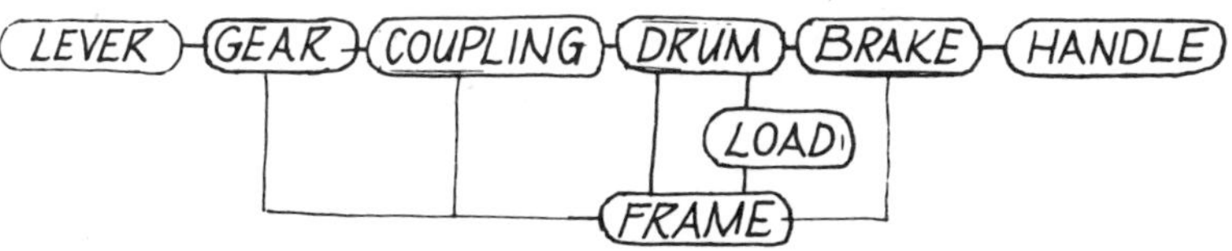

WINDLASS TYPE WINCH

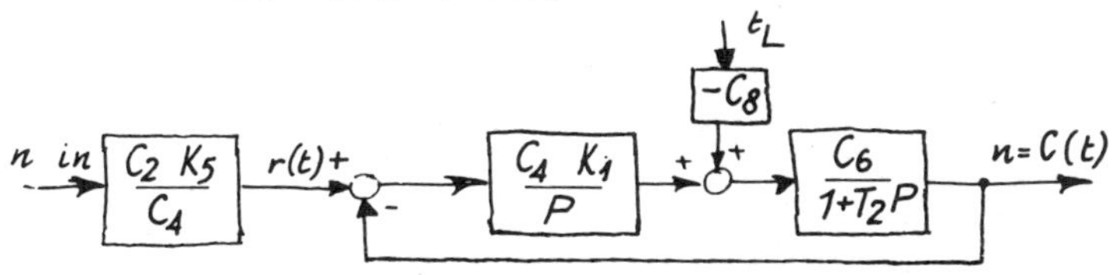

The process in a control system

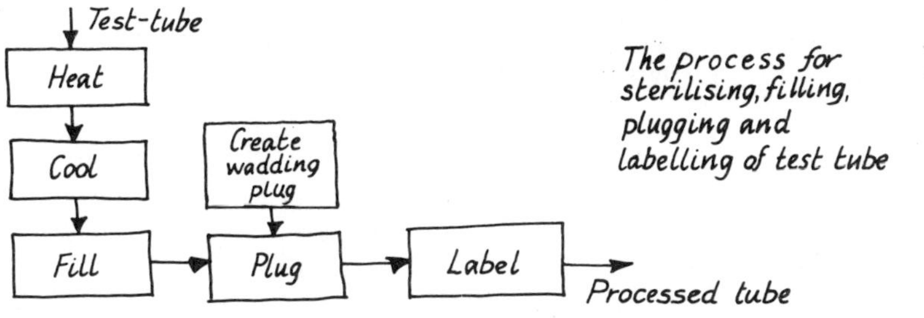

The process for sterilising, filling, plugging and labelling of test tube

DRAWINGS OF PRINCIPLE

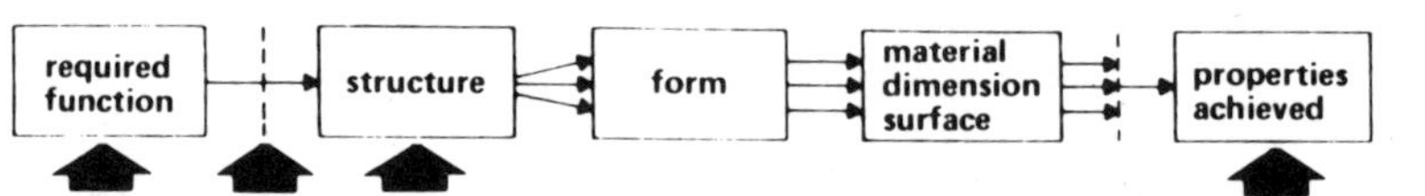

Term

Drawing of principle, diagrammatic drawing.

Use

A drawing of principle is as simple as possible and illustrates the 'principle' or 'operation' of a system. They are used for manipulating ideas and suggestions towards solutions in the syntheses phase of design, and enable the designer to distinguish more clearly between the different possibilities.

Characteristics

Elements are drawn symbolically in order to show the principle more clearly, without being clouded by superfluous details. (Whilst working with principles, presentation of form, dimensions, etc. becomes distracting.)

The symbols can be divided into the following groups:

Mechanical symbols	Page 47
Hydraulic and pneumatic symbols	Page 50
Electrical symbols	Page 51
System symbols	Page 52

Although the groups have been listed separately, there is no reason why they should not be combined in some situations.

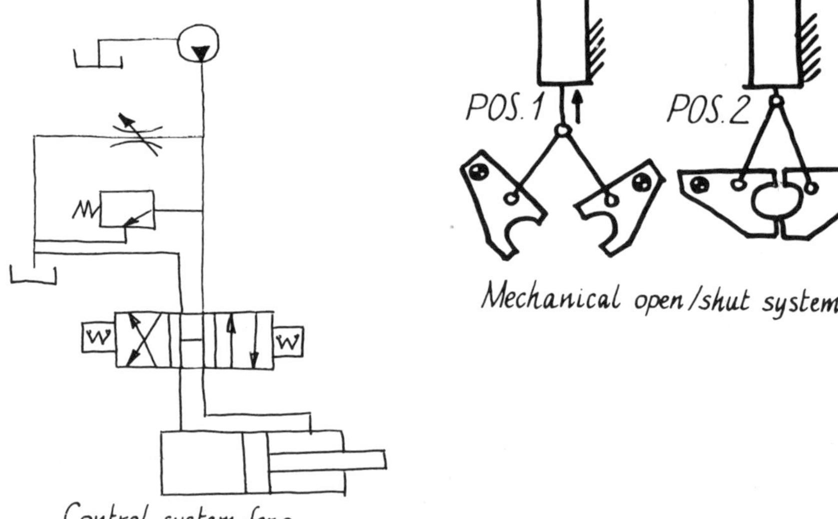

Mechanical open/shut system

Control system for a double-acting hydraulic cylinder

MJE 2955
+15V
1k
2N5193
56Ω
CMC 5LV
VH 248
1μ
320μF/6.4V
1μ
0
16V ~
25000μF 25V

A rectified D.C. power supply

LAYOUTS

Term

Layout

Use

Presentation of form, and relationships between formed elements in detail design, particularly at the stage where the final form and sizes are being decided.

Layouts are used in an 'experimental' way, since forms and sizes are normally decided on the basis of a number of steadily improved layouts.

Characteristics

Layouts are drawn to scale using pencil to facilitate easy correction. It is usual to employ orthographic projection and use a draughting machine. The exception is the manual draughting of ship or aircraft profile lines, however these are now usually generated on a computer and automatically plotted.

The form of the individual elements and the way in which they are assembled, are shown on the layout. For this reason there are benefits in drawing full size to obtain a correct impression of size and spacing.

In addition to the main orthographic views layouts often contain miscellaneous section drawings and detail drawings of individual parts. Layouts may also contain explanatory texts and notes.

The layout is the starting point for detail and assembly drawings.

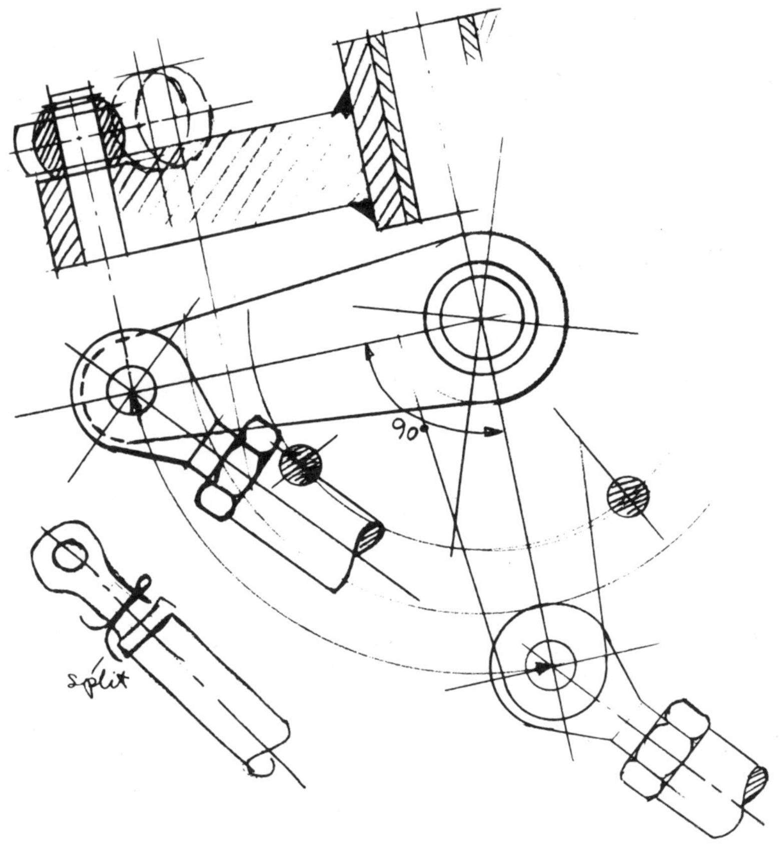

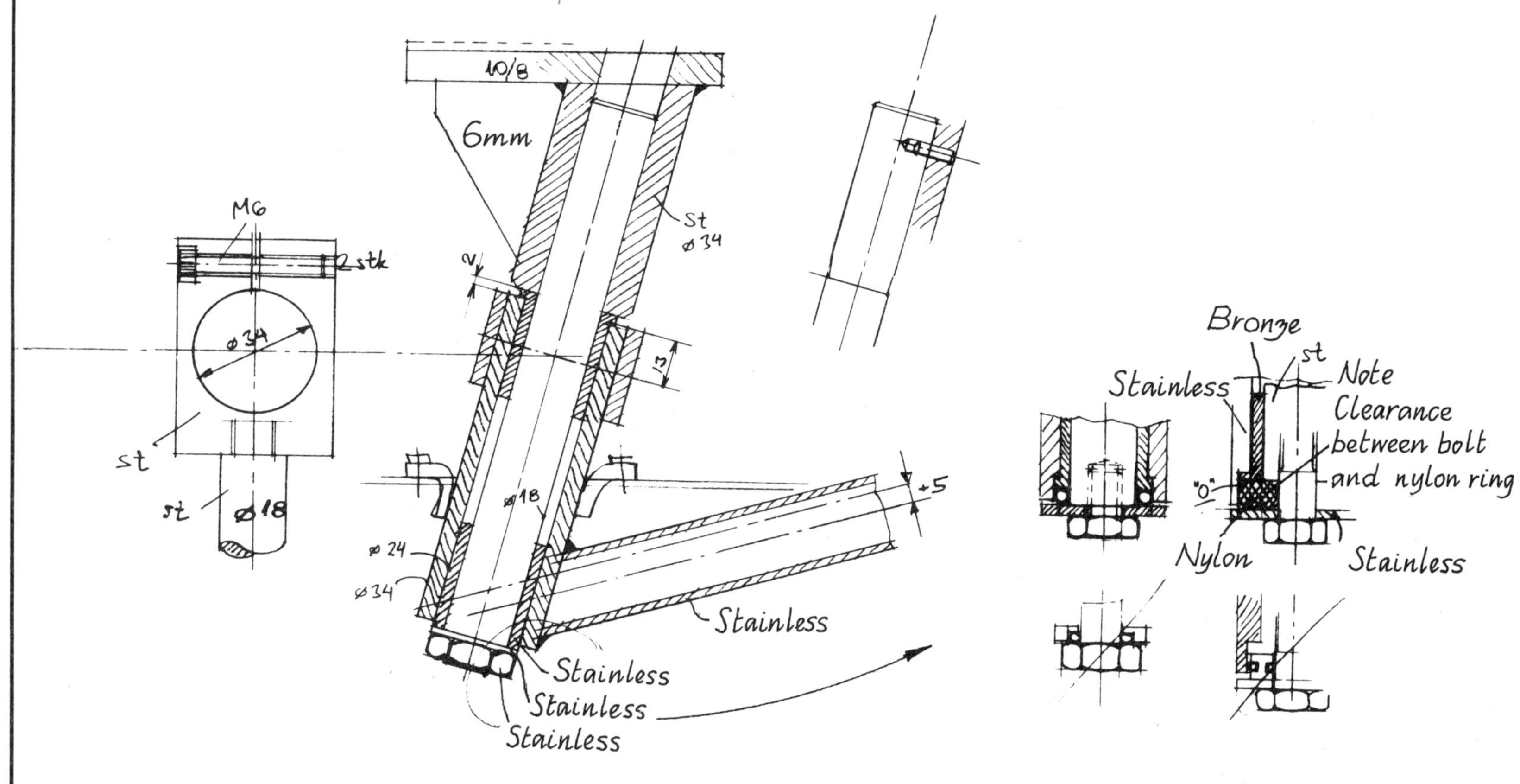
M6
2stk
⌀34
st
st
⌀18
10/8
6mm
st
⌀34
2
3
⌀18
⌀24
⌀34
+5
Stainless
Stainless
Stainless
Stainless
Bronze
st
Stainless
Note
Clearance
between bolt
and nylon ring
"O"
Nylon
Stainless

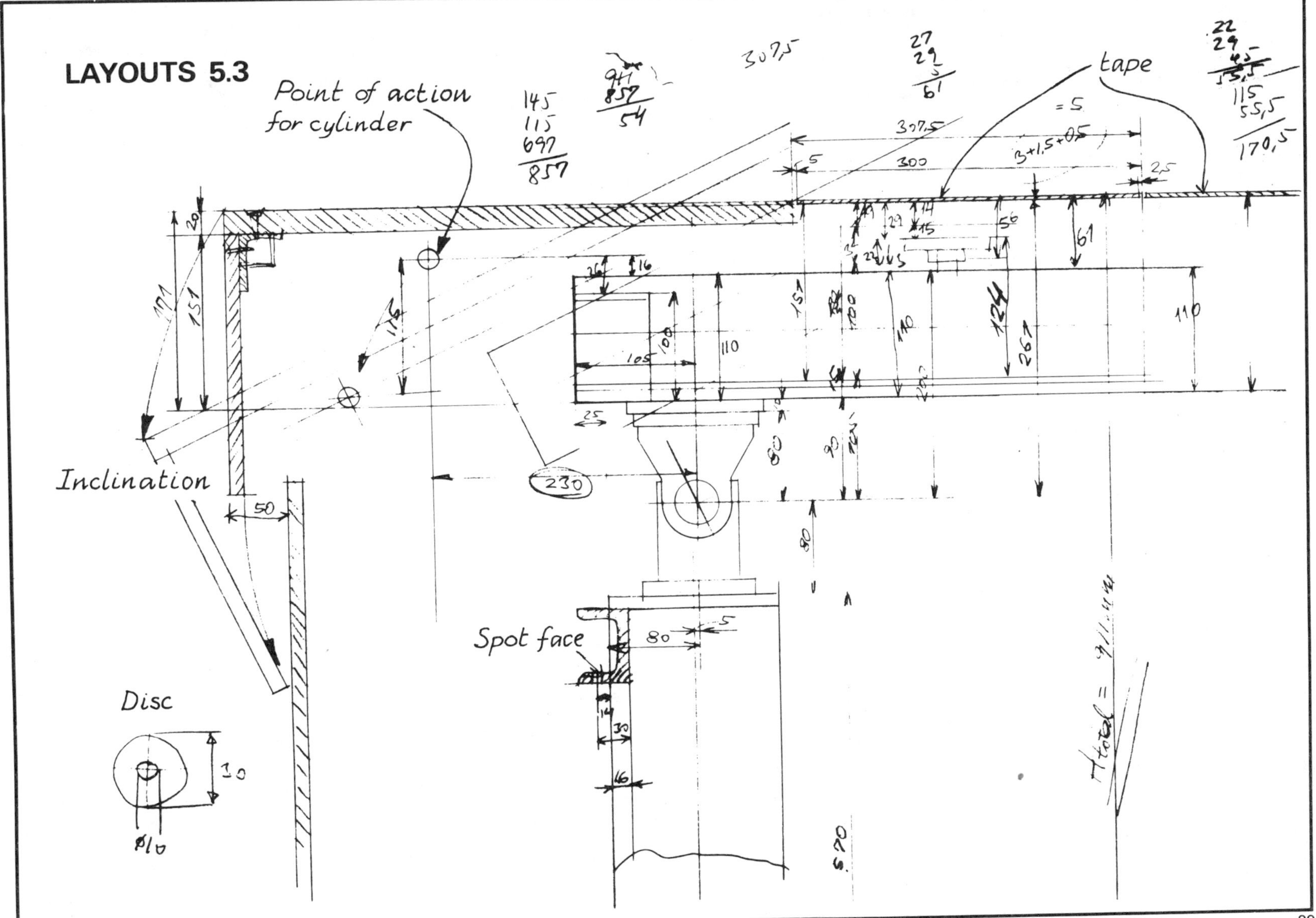
LAYOUTS 5.3
Point of action for cylinder
tape
Inclination
Spot face
Disc
ø10
230
307,5
300
50
80
110
151
171
124
267
61
56
570

5.4 TYPES OF DRAWING

DETAIL DRAWINGS

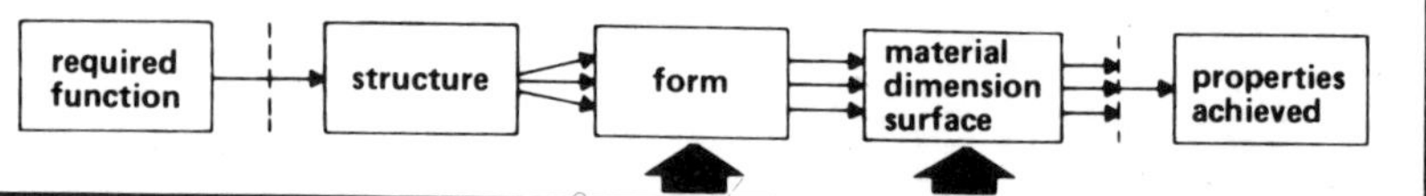

Term

Detail drawing.

Use

Detail drawings specify an object to be produced by the workshop, and consequently are a means of communication between the designer and the workshop. The detail drawing may be drawn by the designer or by a draughtsman. The detail drawing is also the basis for organising production, and for cost assessment.

Characteristics

A detail drawing is usually to scale. It is advantageous to work to one of the common scales e.g.
. . . . . . , 5:1, 2:1, 1:1, 1:2.5, 1:5, 1:10 . . . . . .

On the drawing are given:
- all the necessary dimensions and tolerances
- the surface finish
- material
- number of parts
- scale
- the description of the system or part system to which the group or object belongs
- title of object
- drawing number
- draughtsman's name
- date

The object is normally drawn in orthographic projection, and sections are often used to provide additional information.

Detail drawings may be produced in pencil or ink. Many reproduction methods use the drawing in the same way as a photographic transparency and consequently drawings are normally made on tracing paper or a similar semi-transparent medium.

Some standard drawing conventions are given on pages 53–55.

THIRD ANGLE PROJECTION

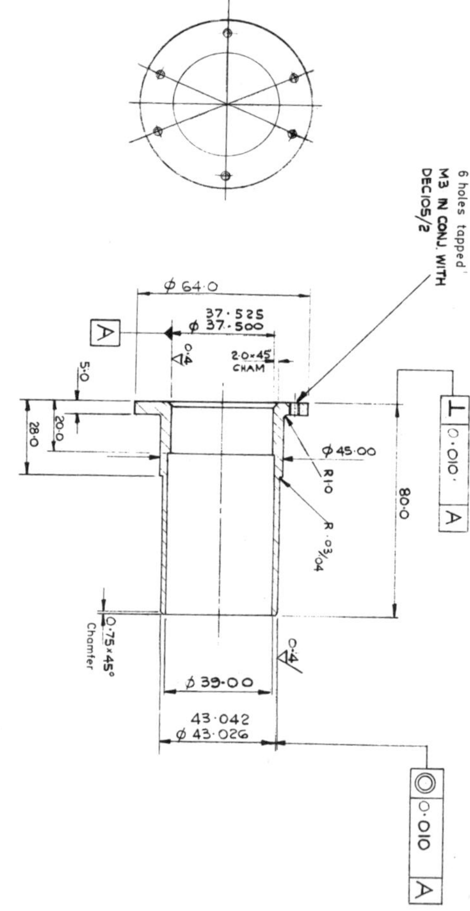

DETAIL DRAWINGS 5.4

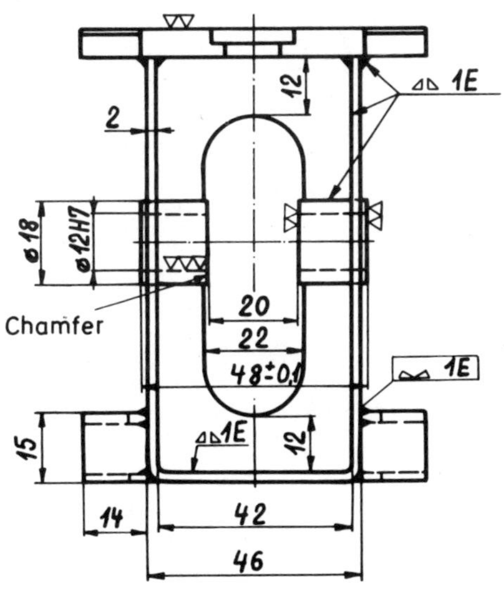

Quantity: 1	Title:	
Mat.: 18/8	Working tolerance: h/H 13	
AMT	Scale 1:1	Sp 6/5-73
Svend H. Pedersen M 60366		
		130705-10

ASSEMBLY DRAWINGS

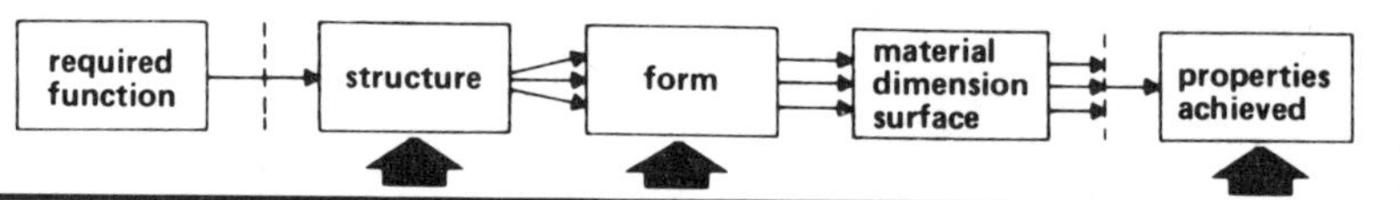

Term

Assembly drawing or general arrangement drawing.

Use

An assembly drawing shows how the individual elements of a product are to be put together i.e. it shows the structure for the finished object.

The assembly drawing includes a *parts list* to aid identification of the individual components.

The assembly drawings and parts lists link the designer's drawings, and provide a ready reference for the designer to find any given element.

The assembly drawing and the parts list are the basis for manufacturing schedules used for programming the manufacture of components in the workshop.

The assembly drawings may be produced by the designer or by a draughtsman.

Characteristics

Assembly drawings are normally to scale. It is advantageous to work to one of the common scales e.g:
. . . . . .10:1, 5:1, 2:1, 1:1, 1:2.5, 1:5, 1:10 . . . . . .
An identification number is given to each element on the drawing, referring to the parts list. The following should also be given:

- the quantity per assembly of each part
- scale
- project title
- title of the assembly
- drawing number
- date
- draughtsman's name

Orthographic projection is normally used, the number of views and sections being chosen according to need.

Assembly drawings may be in pencil or ink. Many reproduction methods use the drawing in the same way as a photographic transparency and consequently drawings are normally produced on tracing paper or a similar semi-transparent medium.

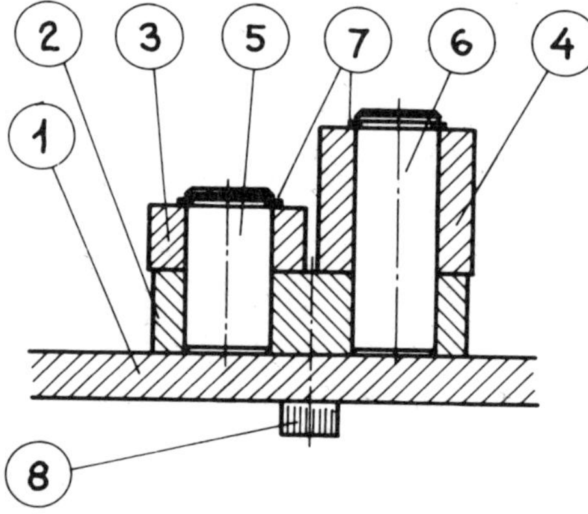

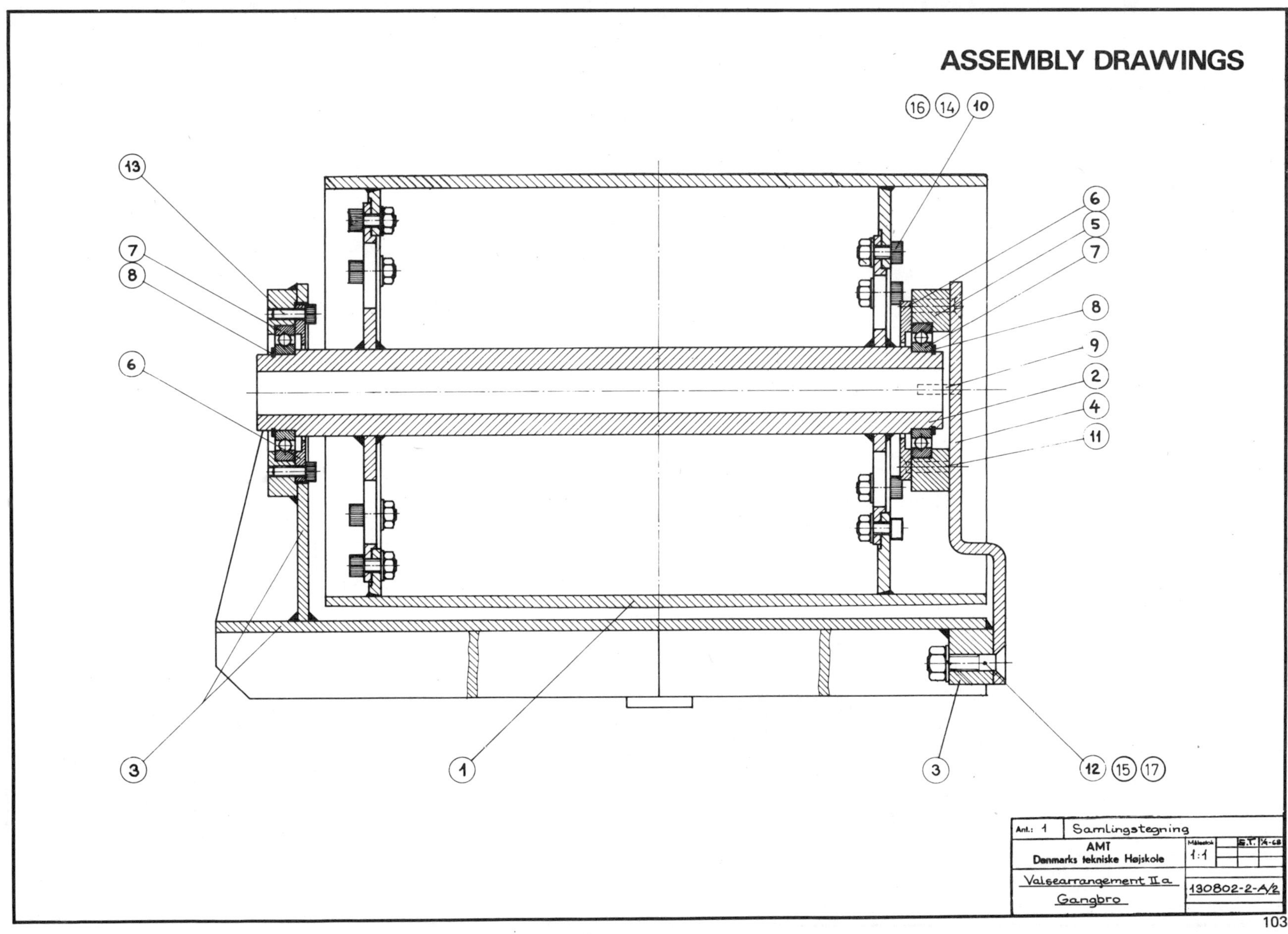
16
14
10
13
6
5
7
8
9
2
4
11
7
8
6
3
1
3
12
15
17
Ant.: 1
Samlingstegning
AMT
Danmarks tekniske Højskole
1:1
Valsearrangement II a
Gangbro
130802-2-A/2

ASSEMBLY DRAWINGS

(Examples of a parts list)

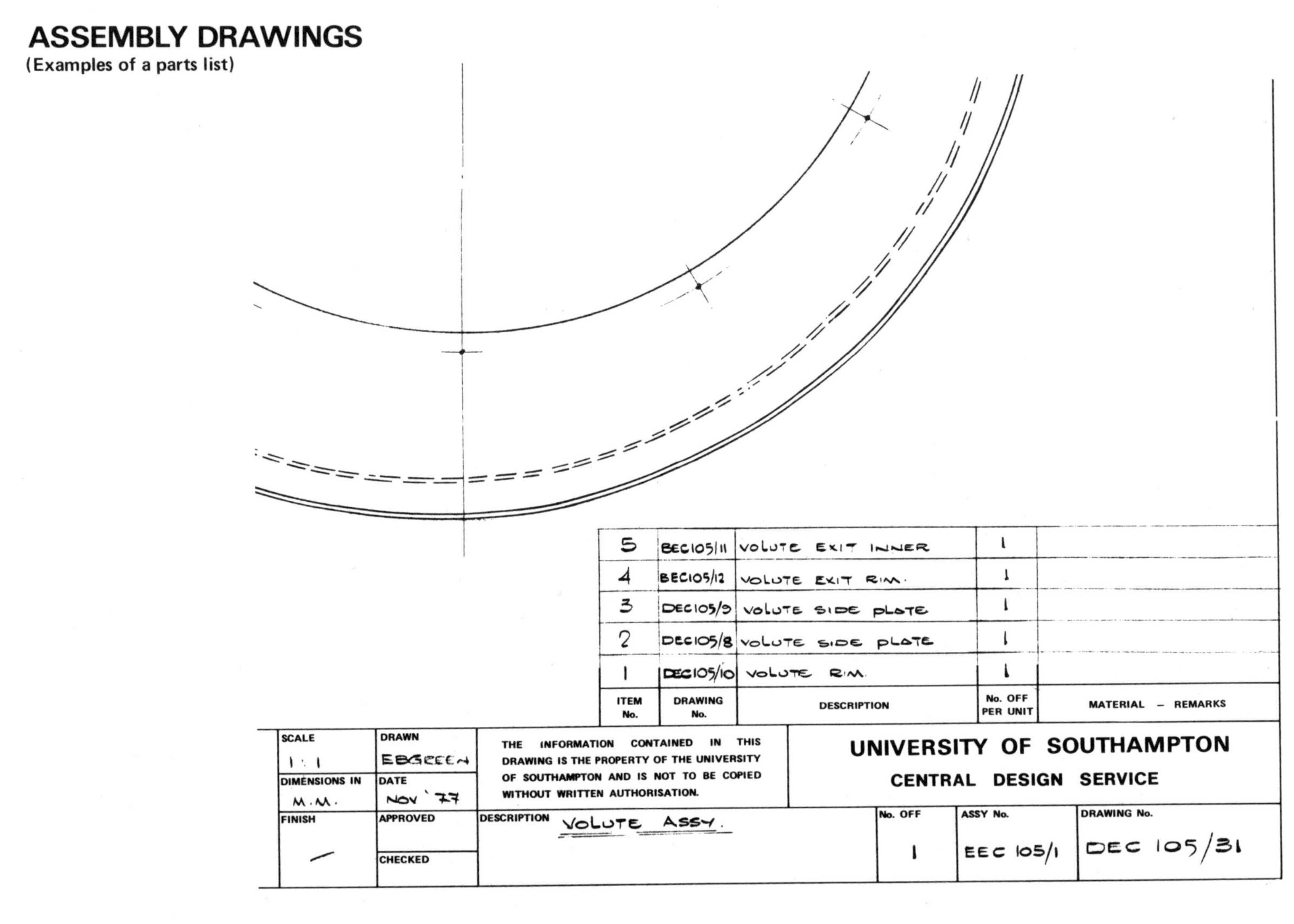

ITEM No.	DRAWING No.	DESCRIPTION	No. OFF PER UNIT	MATERIAL – REMARKS
5	BEC105/11	VOLUTE EXIT INNER	1	
4	BEC105/12	VOLUTE EXIT RIM.	1	
3	DEC105/9	VOLUTE SIDE PLATE	1	
2	DEC105/8	VOLUTE SIDE PLATE	1	
1	DEC105/10	VOLUTE RIM.	1	

EXPLODED VIEWS

Term

Exploded view, exploded drawing.

Use

To represent a number of mechanical elements, and their mutual relationships, i.e. a description of the arrangement of a finally designed system. These drawings are normally used for describing repairs and spare parts.

Characteristics

As far as possible all the parts are shown separately as though the parts were being "pulled out" along the axes or the directions in which they are assembled. Sometimes these axes are drawn, to make it clear where the parts belong.

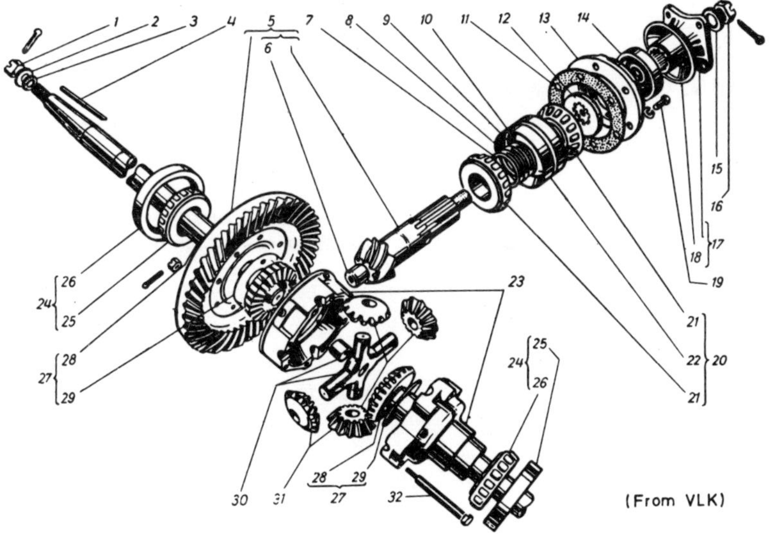

(From VLK)

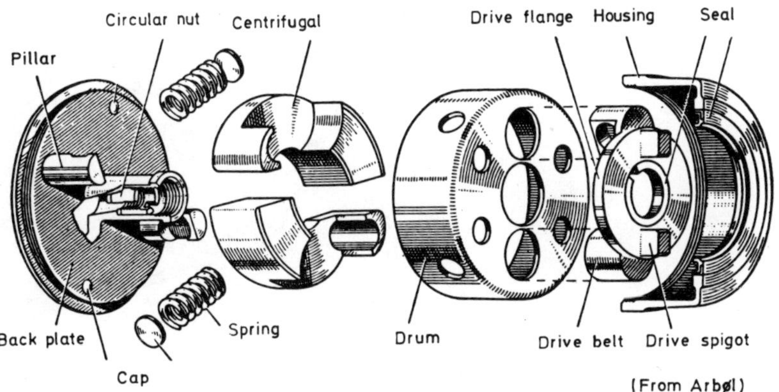

(From Arbøl)

OVERLAY DRAWINGS

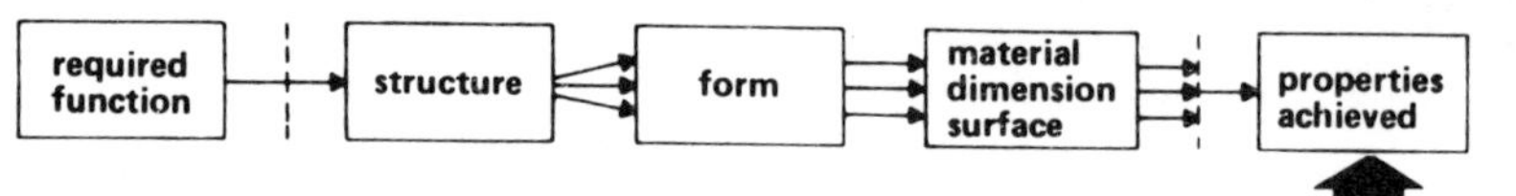

Term

Overlay drawing, X-ray drawing.

Use

Drawing of items so that hidden areas and parts can also be seen. In effect the inner and outer forms are superimposed on each other so that their relative positions are evident.

Characteristics

Surfaces hiding parts to be drawn are treated as 'transparent', by showing only the contours.

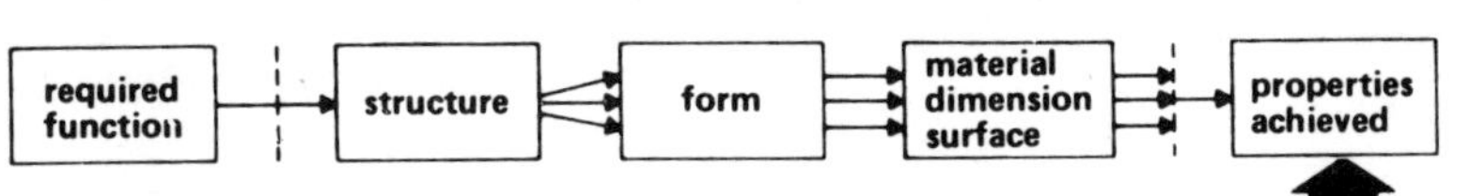

PATENT DRAWINGS

Term

Patent drawing.

Use

Illustration of the system or method covered by the patent. It is an integral part of the patent description.

Characteristics

Patent drawings tend to follow a particular convention as shown in the example.

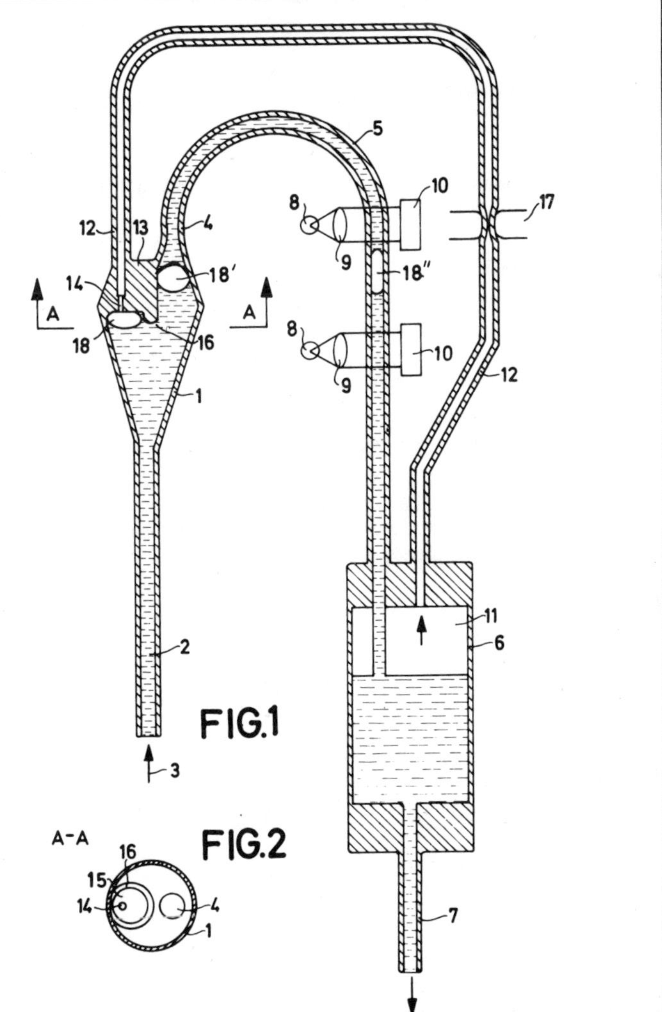

BAR OR COLUMN DIAGRAMS

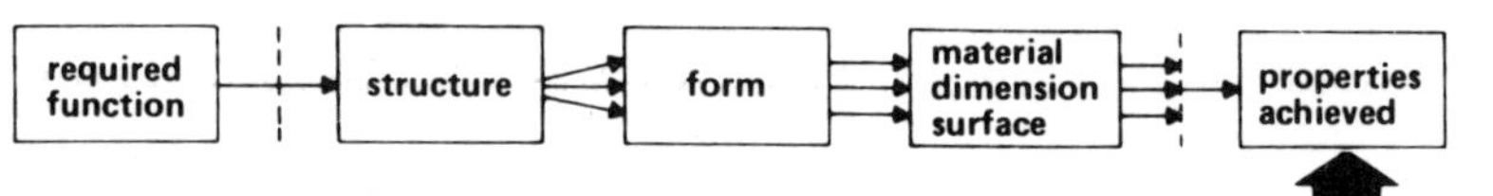

Term

Column diagram, bar chart, histogram.

Use

A convenient way of showing, with impact, the dependence of one or more parameters on another parameter.

Characteristics

The basis for a column diagram is a Cartesian co-ordinate system in which one of the co-ordinate values increases in steps, whilst the other is shown as columns of specific lengths. The co-ordinates may be completely or partly omitted.

Bar chart is the term normally applied to the various forms where the 'columns' are displayed horizontally.

The term histogram is used if the columns touch each other so that they appear as a stepped curve, and where both parameters are represented numerically. The histogram only represents the relationship between two parameters, one on each axis.

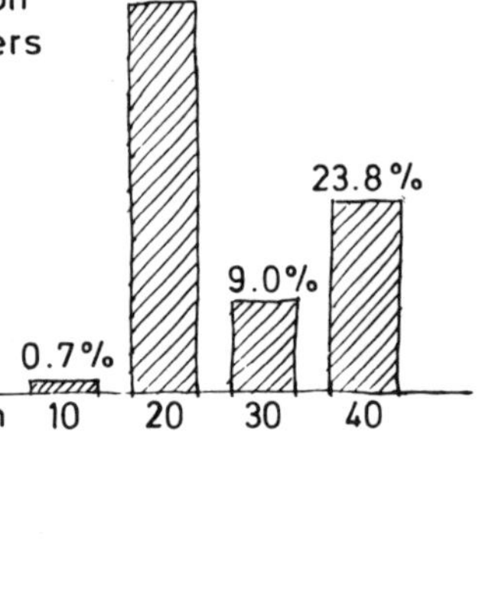

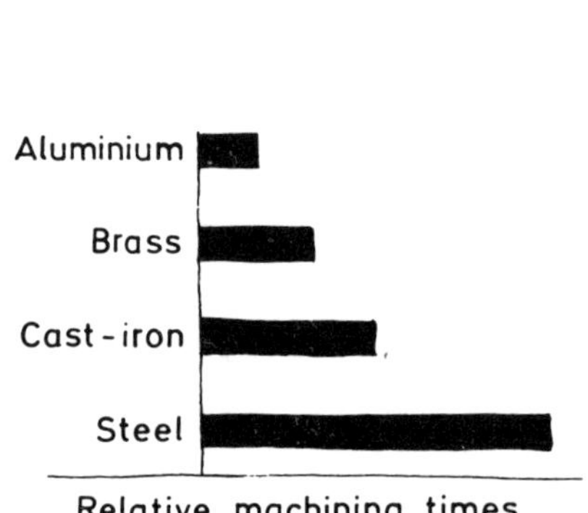

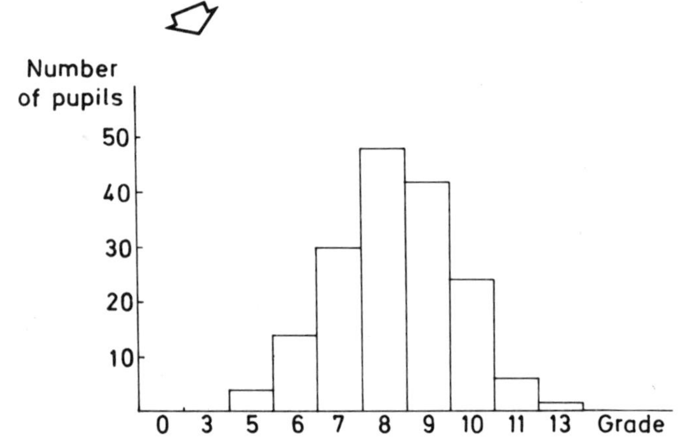

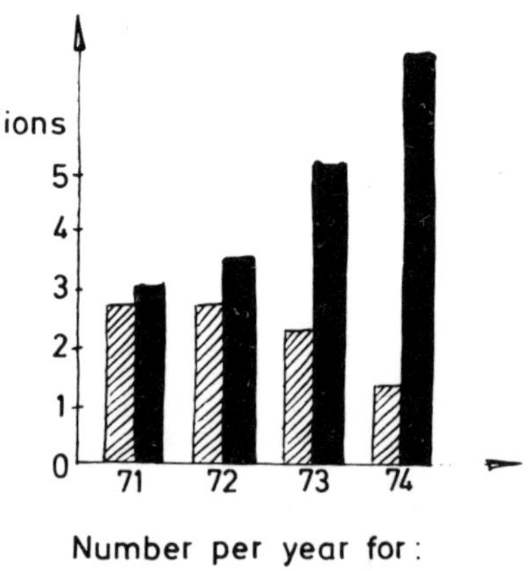

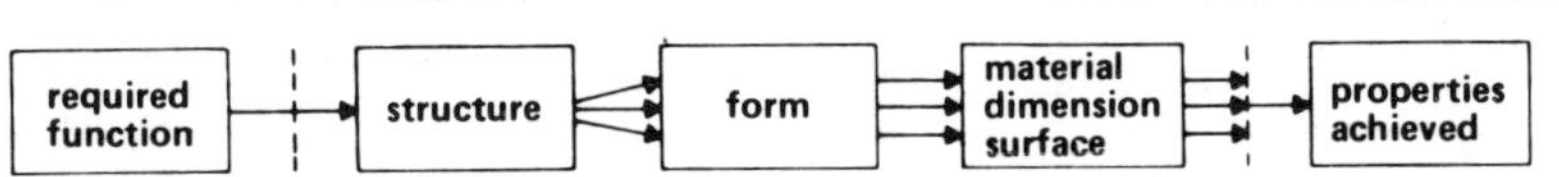

GRAPHS (Cartesian co-ordinate system)

Term

Graph, curve, plot.

Use

For graphically relating two parameters, one being a function of the other.

Characteristics

The coordinate axes are calibrated at set intervals. The calibration is generally linear or logarithmic, and the axes may be to different scales.

% of body weight
30
20
10
-10
-20
-30
10 20 30 40 50 % of period

Axial force

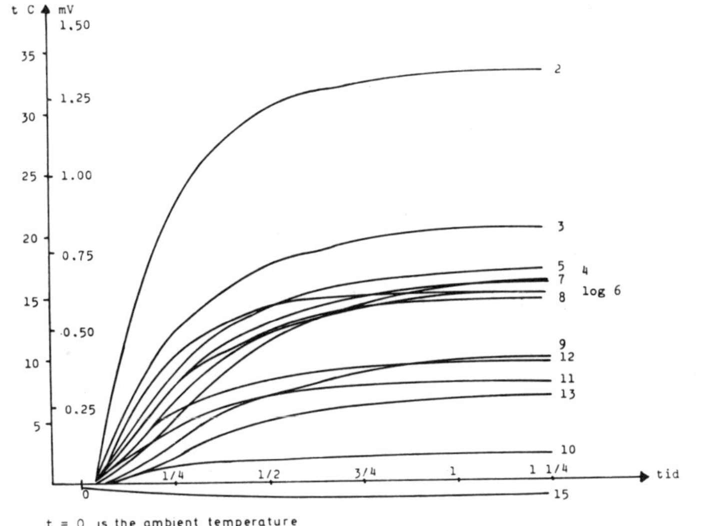

t = 0 is the ambient temperature

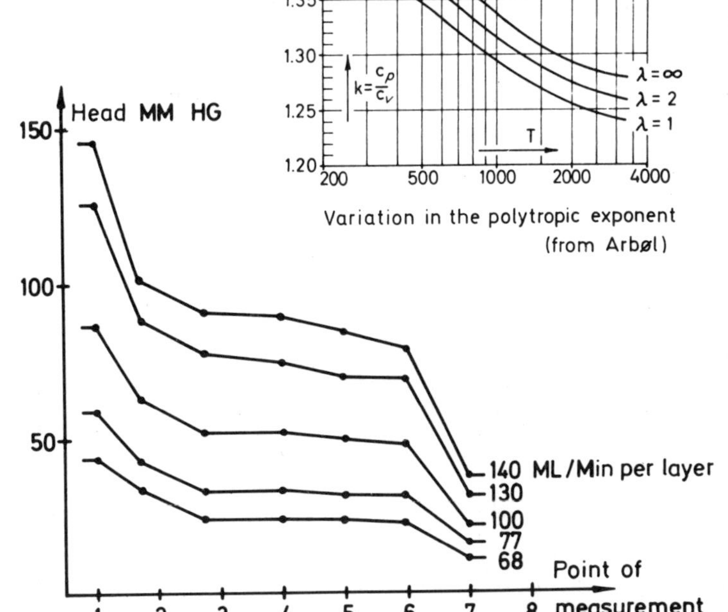

Variation in the polytropic exponent (from Arbøl)

GRAPHS (Polar co-ordinate system)

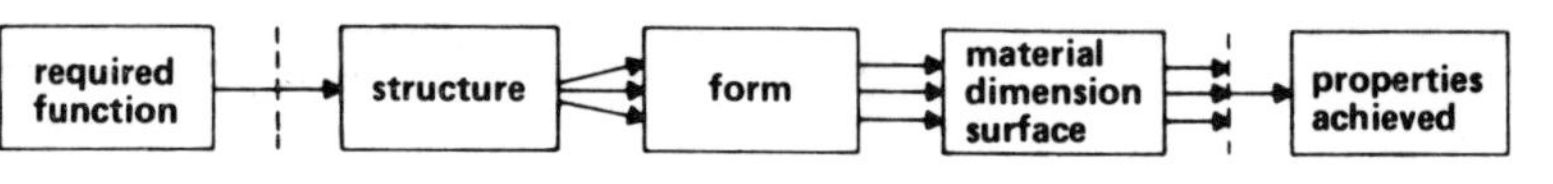

Term

Polar plot, graph in polar co-ordinates.

Use

A drawing of the relationship between two parameters, one being a function of the other. Normally used where a value can be related to an angular position or where a relationship is represented by complex variables on an argand diagram.

Characteristics

Any point on the curve is decided by a direction from the zero line (the polar axis), and a distance from a zero point (the pole).

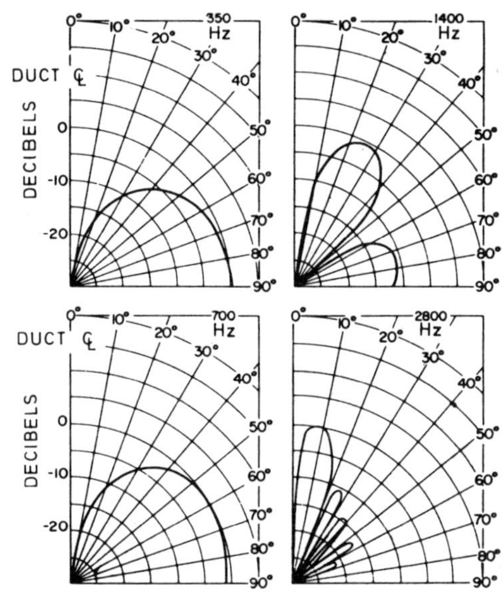

[FROM HOELSCHER]

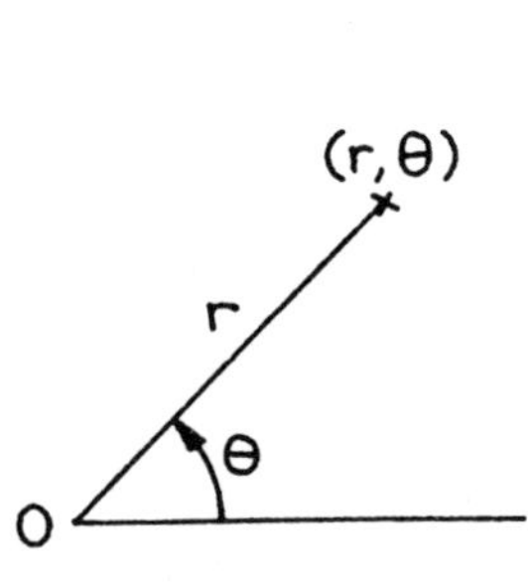

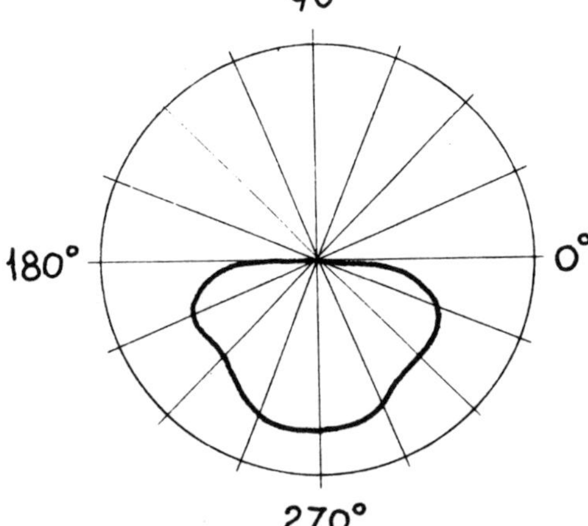

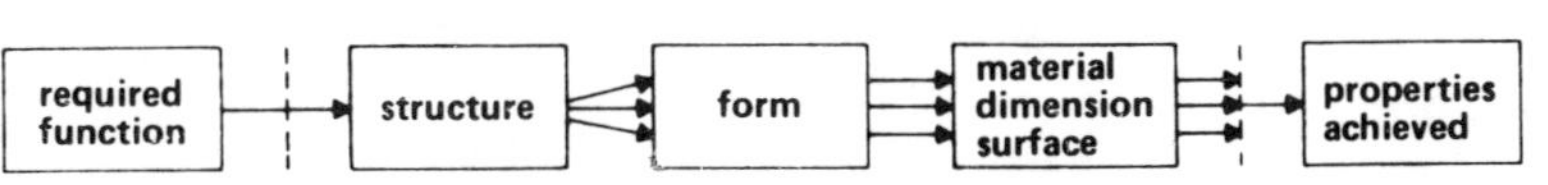

REPRESENTATION OF SURFACES

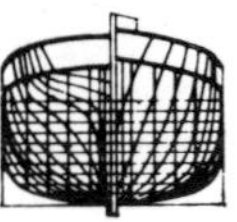

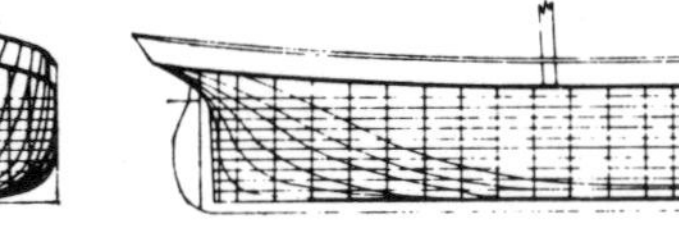

Term

Surface representation, nested sections.

Use

Drawing of the relationship between three parameters.

Characteristics

Surfaces are conveniently represented by drawing them as a net, or by the surface being cut by a number of equispaced parallel planes or by planes set at angular intervals.

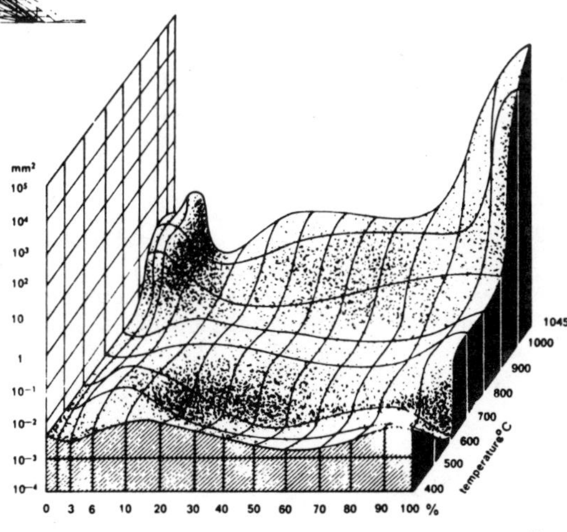

[FROM A. ALMAR-NAESS]

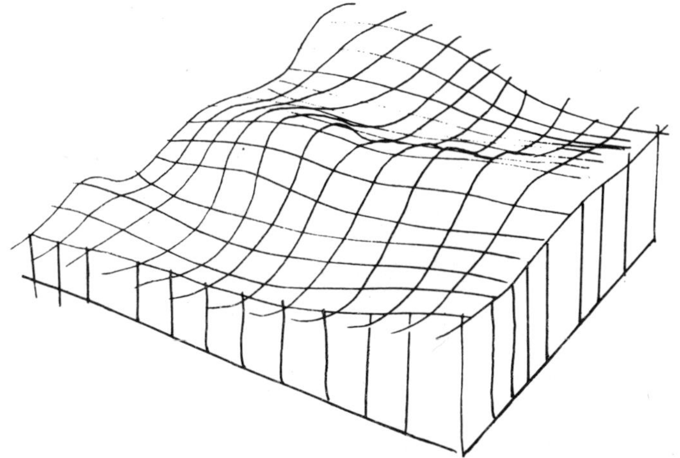

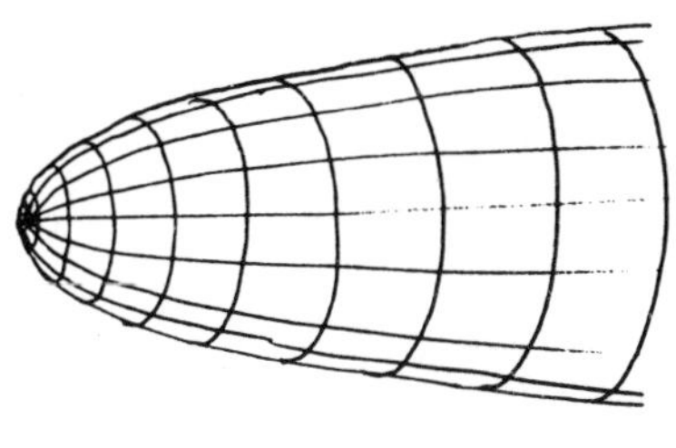

5.13 TYPES OF DRAWING

TRILINEAR CHARTS

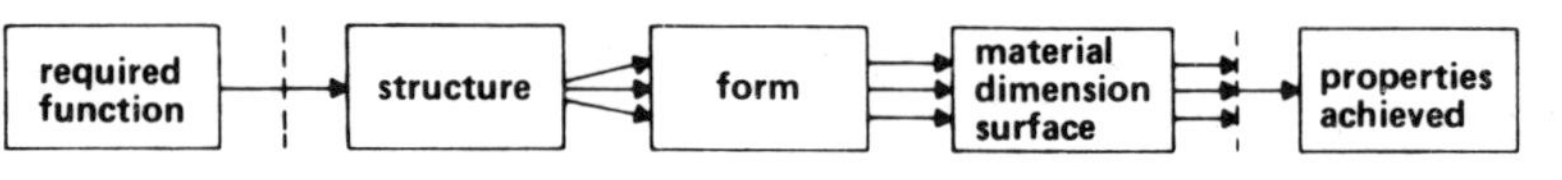

Term

Trilinear diagram, trilinear chart.

Use

For relating three parameters any one of which is the compliment of the other two. Used for representing the properties of alloys.

Characteristics

The diagram appears as an equilateral triangle having a number of lines parallel with each side. A given line represents a certain percentage of one of the parameters. A specific point in the diagram then gives the percentage of the three parameters.

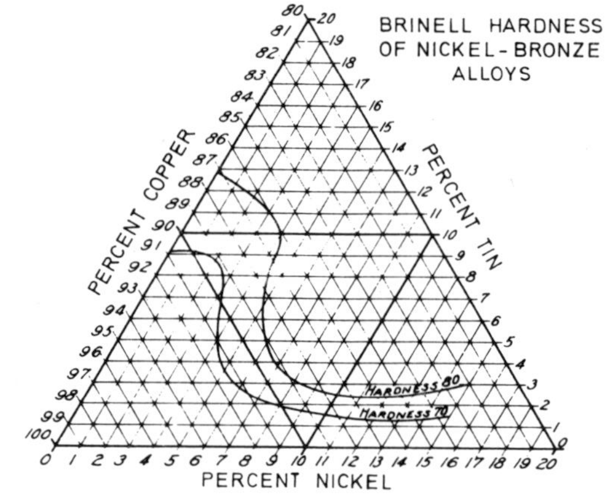

[FROM HOELSCHER]

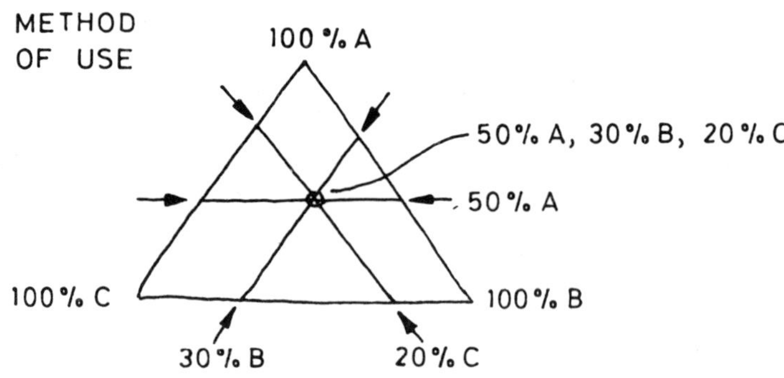

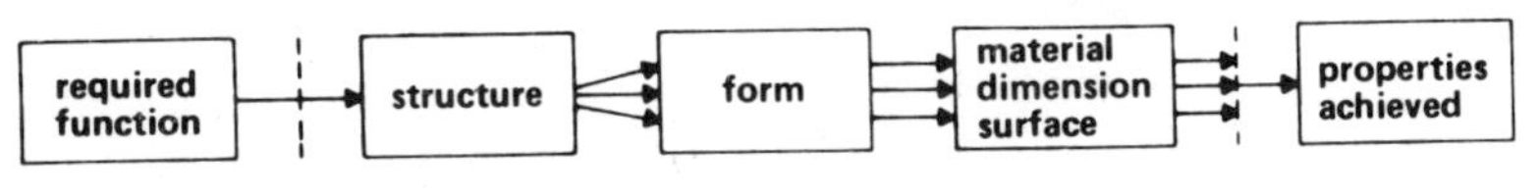

DIAGRAMS

Term

Diagram

Use

For representing the relationship between three or more parameters.

Characteristics

Such diagrams need supplementary text, as only the connection between two variables can be drawn in a single plane figure. Where the parameters, other than those represented by the two coordinates, can take on more than one value, the diagram must be more specifically defined.

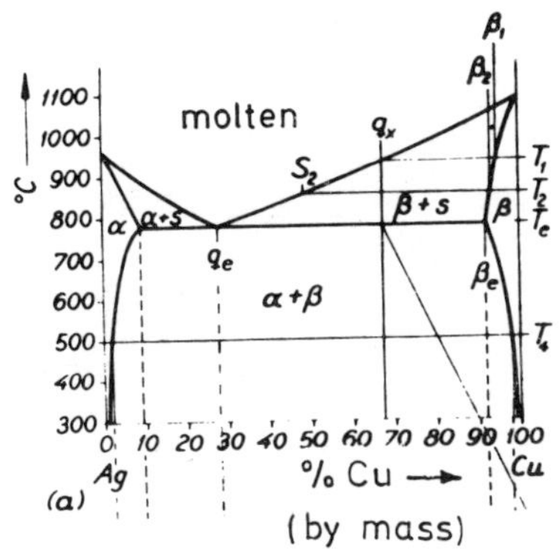

(From A. Almar-Naess)

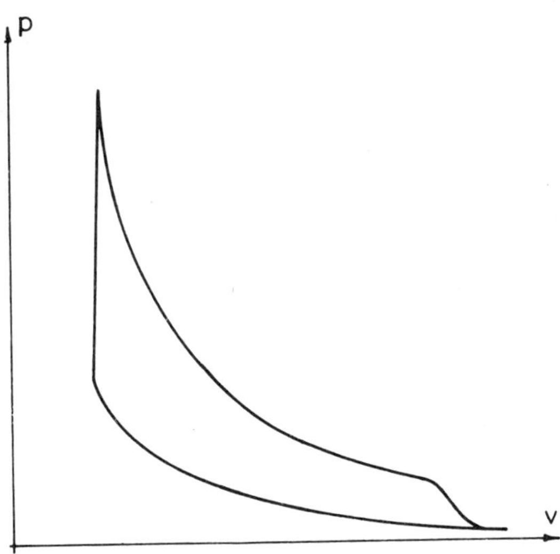

Pressure-volume diagram for a two stroke petrol engine

(From Arbøl)

SECTOR DIAGRAMS

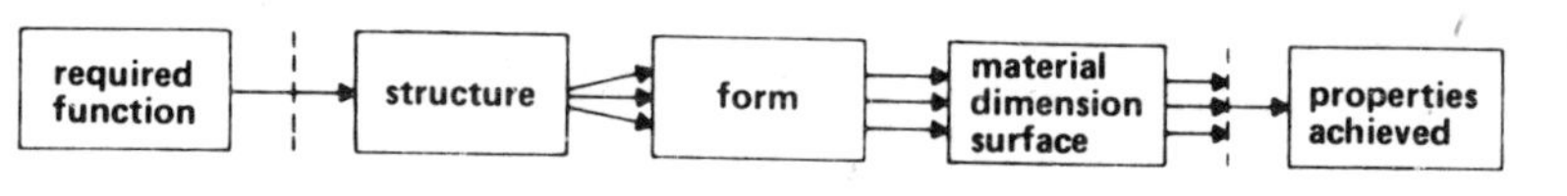

Term

Sector diagram, pie chart, slice of the cake diagram

Characteristics

A circle representing 100% is subdivided into sectors to show the proportion attributable to the aspect named on a given sector. There may be no calibrated scale, the diagram relies on the eye's ability to readily divide a circle into parts.

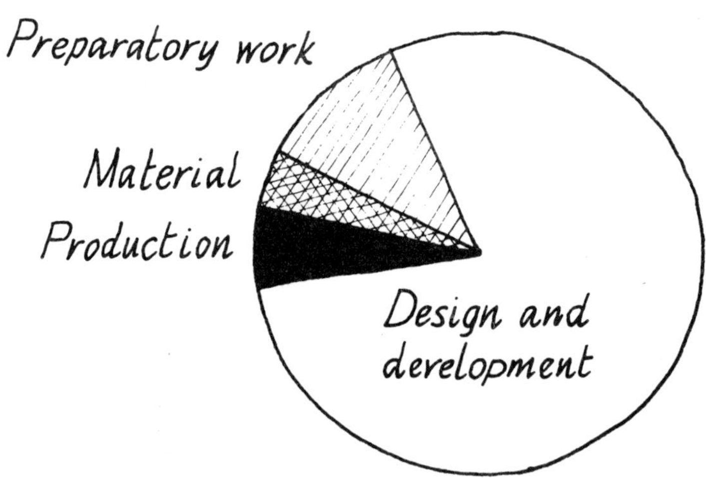

Example showing direct costs for a product

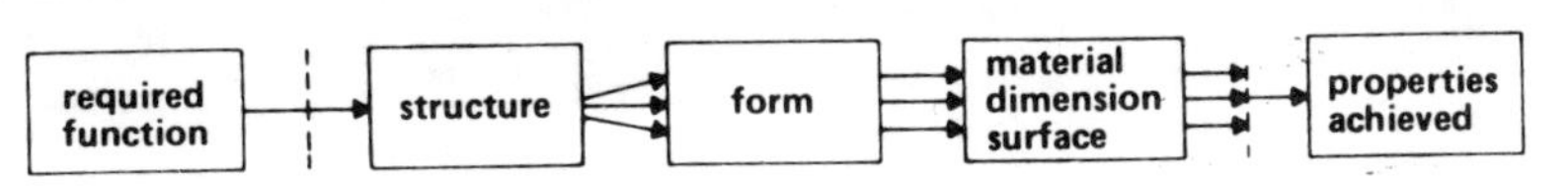

PERCENTAGE COLUMN

Term

Percentage column, percentage bar, 100% column, 100% bar.

Use

A representation of a total quantity and the way it is divided into its constituent parts.

Characteristics

The diagram comprises a column divided into areas corresponding to the size of the components. The areas are sometimes coloured, and or hatched, etc.

0.7% 10′
9.0% 30′
23.8% 40′
66.5% 20′

5.17 TYPES OF DRAWING

NOMOGRAM

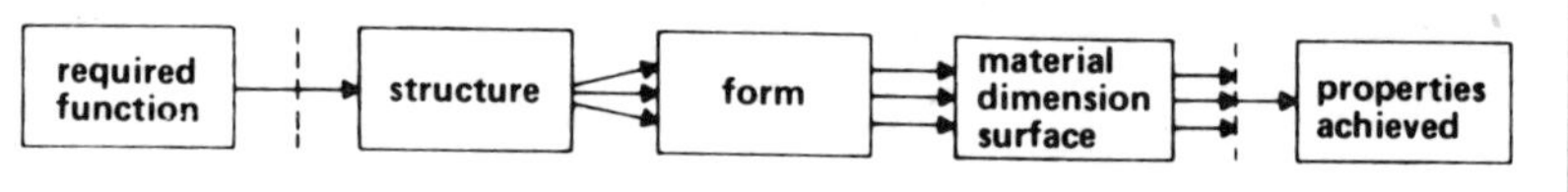

Term

Nomogram

Use

Nomograms are used where the solutions to an equation are required for several different data values. The nomogram in effect represents all the solutions to the equation. These solutions are revealed when one or more straight lines are constructed through the data points on the scale where the corresponding solution is required.

Characteristics

Nomograms must have at least two calibrated axes. They can be constructed in many ways. The examples on the right show a limited selection.

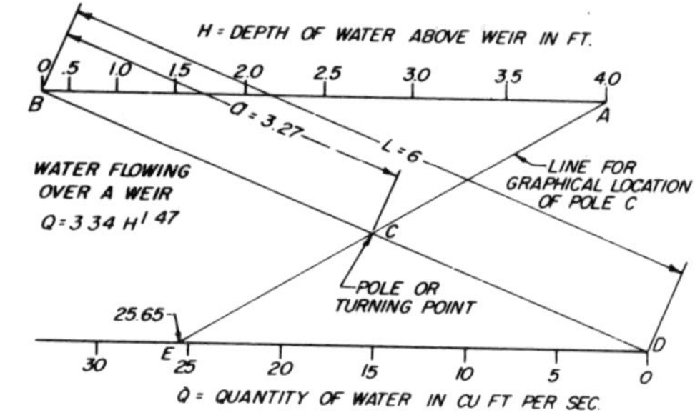

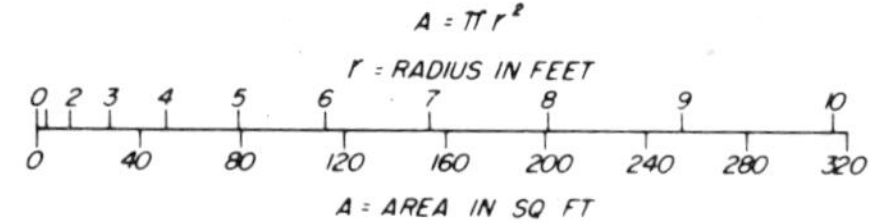

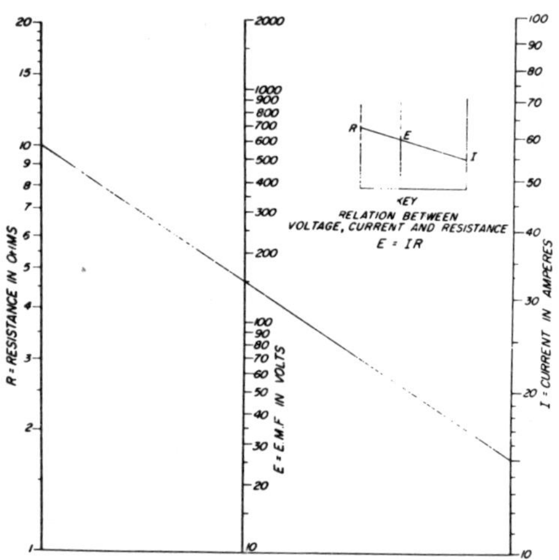

[FROM HOELSCHER]

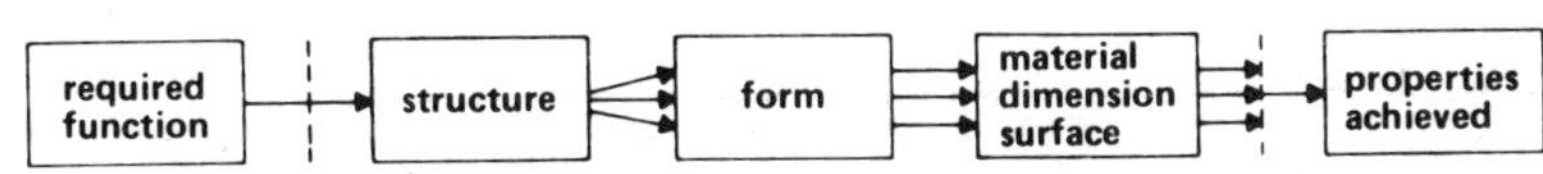

OTHER TYPES OF DRAWING

Common Unnamed Drawing Types
Many design drawings are of unnamed types. The term "freehand sketch" is often applied to drawings of form expressed as a projection but produced freehand. Other unnamed drawing types can be found on the procedure sheets in group 1 referring to "Modelled properties".

This important group should not be overlooked. Two frequently used categories are represented by examples shown on this page.

Drawing characteristics (see page 5):–
FORM–SELF COMMUNICATION–PROJECTION–FREEHAND SKETCHING
Suggested form for a chromosome analysis apparatus.

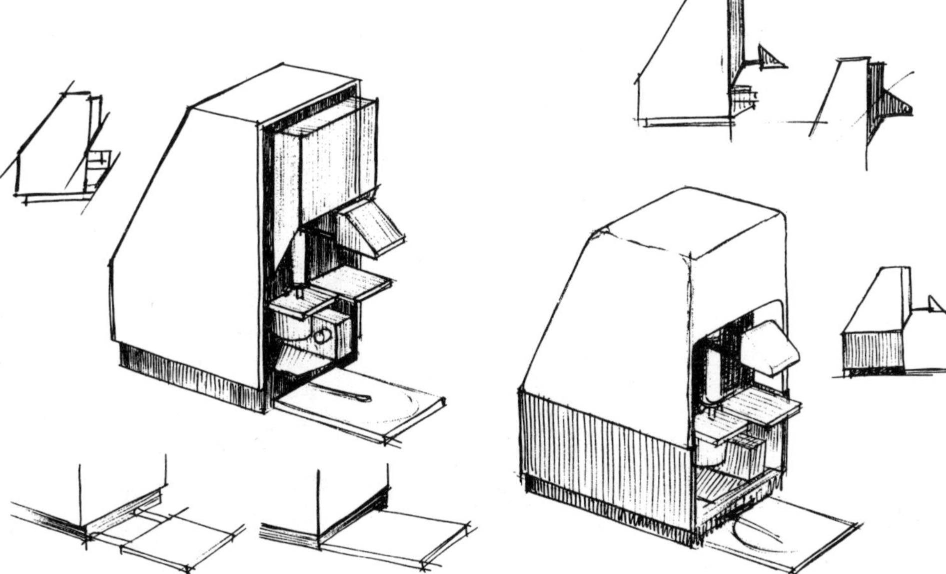

Drawing Characteristics:–
FUNCTION – SELF COMMUNICATION – SYMBOLS – FREEHAND SKETCHING
Alternative arrangements for winding a rectangular electric coil.

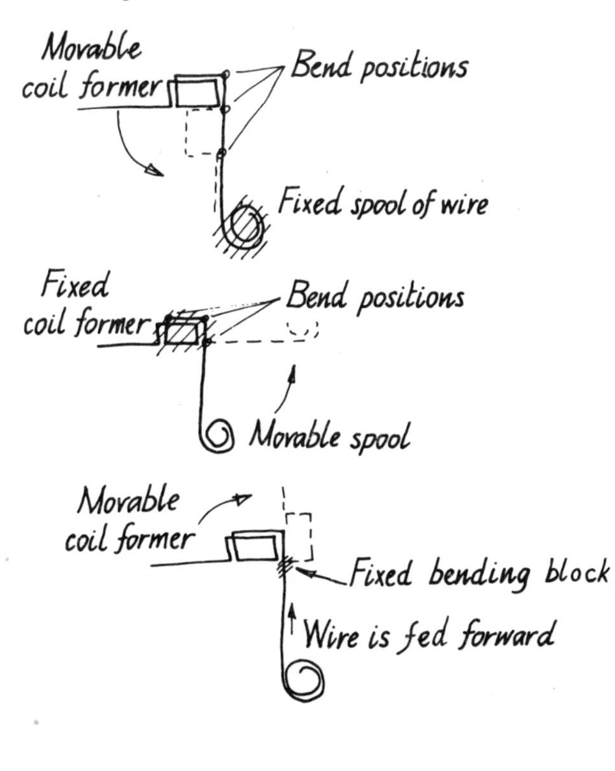

EVOLVING NEW DRAWING TYPES – SOME EXAMPLES

It is possible to develop new drawing types by establishing the level at which the drawn property is required, this also decides the amount of information required of the drawing. New symbols may also be created for the purpose. Below are two examples. (N.B. it is useful to define the symbols where they are new symbols or symbols used in an unusual context.)

Drawing characteristics
STRUCTURE – SELF COMMUNICATION–SYMBOLS–FREEHAND SKETCHING

Suggestions for different arrangements of the components of a road-roller.

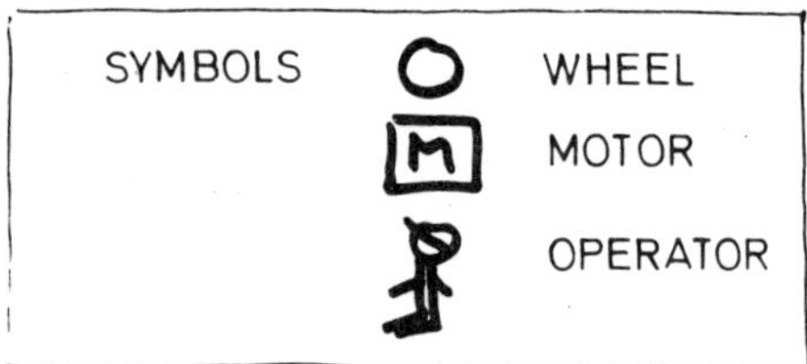

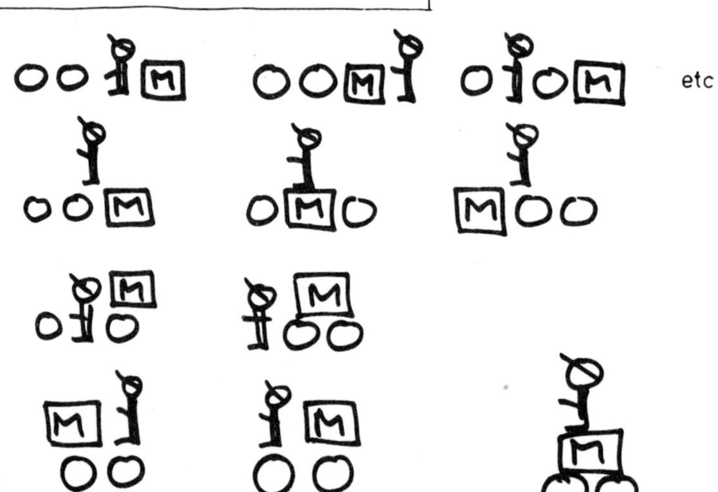

OTHER TYPES OF DRAWING

Drawing characteristics
FORM-SELF COMMUNICATION–SYMBOLS–FREEHAND SKETCHING

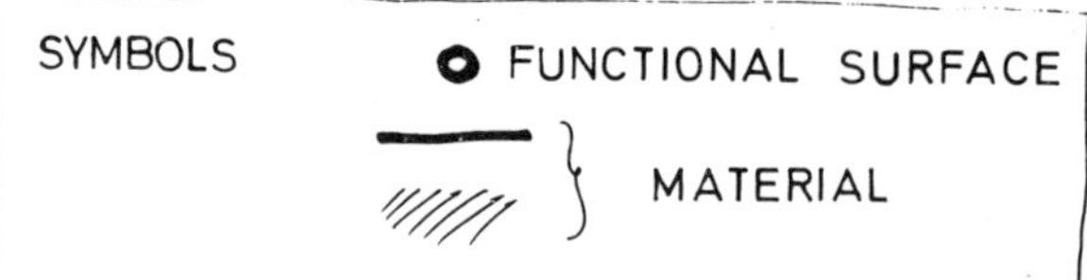

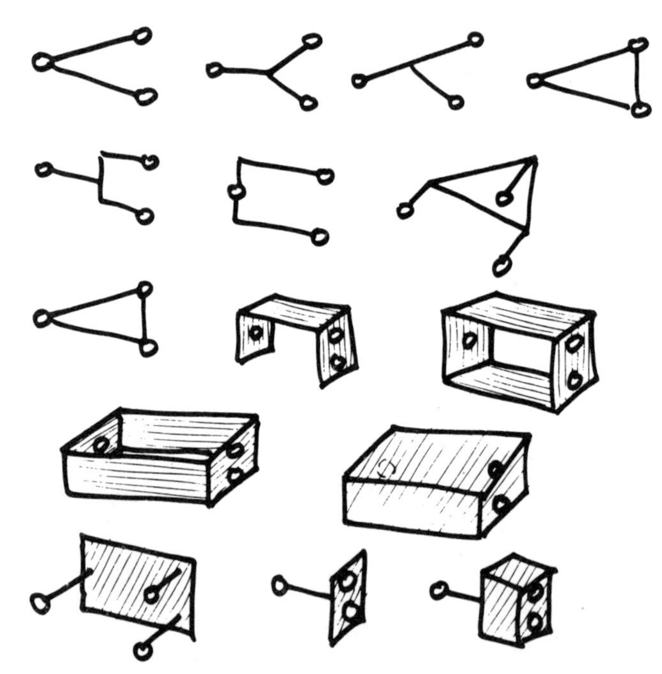

Bibliography

Books

Cooley, *Engineering Drawing, Communication & Design*, Pitman (1974)
Dunne, P. M., *Engineering Graphics*, Pergamon Press (1967)
Giesecke, F. E. *et al., Engineering Graphics*, Macmillan (1969)
Hoelscher, R. P. *et al., Graphics for Engineers: visualization, Communication, Design*, John Wiley & Sons (1968)
Hoischen, H., *Technisches Zeichnen*. Grundlagen, Norman, Beispiele, Verlag W. Girardet, Essen (1970)
Hubka, V., *Theorie des Konstruktionsprozesses,* Springer-Verlag, Germany (1966)
Krick, Edward V., *An Introduction to Engineering and Engineering Design,* John Wiley & Sons (1969)
Levens, A. S., *Graphics, Analysis and Conceptual Design*, John Wiley & Sons (1968)
Oldham, E. A. and Mee, C. W., *Engineering Drawing and Materials*, Intertext Books (1968)
Pitts, G., *Techniques in Engineering Design,* Newnes-Butterworths, London (1973)
Pocock, K. A., *Engineering Drawing and Graphics*, Newnes-Butterworths, London (1972)
Reimpell, J. m.fl., *Die normgerichte technische Zeichnung für Konstruktion und Fertigung*. Band 1 u. 2., DVI-Verlag GmbH, Düsseldorf (1967)
Tjalve, *A short course in industrial design*, Newnes-Butterworths, London (1979)

British Standards

BS 203: 1978. *Engineering Drawing Practice*
BS 3939. *Graphical Symbols for Electrical Power, Telecommunications and Electronics Diagrams*
BS 4058: 1973. *Data Processing Flow Chart Symbols, Rules and Conventions*
BS 4500. *ISO Limits and Fits*
BS 5070: 1974. *Drawing Practice for Engineering Diagrams*